Radionuclide tracer techniques in haematology

Radionuclide tracer techniques in haematology

C. S. Bowring, BSc, PhD, MInstP

Principal Physicist, Department of Physics, Royal Devon and Exeter Hospital, Exeter

Butterworths

London Boston Sydney Wellington Durban Toronto

First published 1981

© Butterworth & Co (Publishers) Ltd 1981

British Library Cataloguing in Publication Data

Bowring, C. S.
  Radionuclide tracer techniques in haematology.
  1. Radioisotopes in haematology
  I. Title
  616.07'561   RB45   80-41804

ISBN 0-407-00183-2

Typeset by Butterworths Litho Preparation Department
Printed in Great Britain by Hartnoll Print Limited, Bodmin, Cornwall

Preface

The use of radionuclides in diagnostic medicine and in haematology in particular does not have a long history. The first appearance in the published literature of the application of radionuclides as an aid to haematological studies was in 1939 when the use of radioactive iron to study iron metabolism was described. It was not until the early 1950s, however, that these, and other studies using radionuclides, really became at all widely used. Since these early days the use of radionuclides has spread into many aspects of haematology and the International Committee for Standardization in Haematology (ICSH) has made several recommendations relating to the more common radionuclide procedures. There is at present, however, no concise guide to the diagnostic uses of radionuclides in this field of medicine and it is necessary for the user to refer to numerous different articles and papers in the textbooks and journals. It is hoped that this little book will help fill this gap; first, by bringing together descriptions of all the commonly used *in vivo* radionuclide techniques in haematology; secondly, by providing information on areas of haematology where the use of radionuclides is still developing; and thirdly, by giving practical instruction on the application of these techniques. Although the book is primarily concerned with the *in vivo* applications of radionuclides, the use of radioassay techniques is briefly discussed in Appendix 4.

It is hoped that this book will be of value to the scientific and technical staff responsible for doing the measurements, regardless of whether they are in departments of haematology, nuclear medicine or medical physics, and also to medical staff who have not previously been involved in the diagnostic uses of radionuclides in haematology. It is accepted that each of these three categories of reader will be looking for something slightly different from the book, and it is hoped that a reasonable balance has been achieved. In particular, the first two chapters, covering the basic physics of radionuclides and the instrumentation used in their detection, may be omitted by readers with experience of nuclear medicine techniques.

I would like to thank many people for the help and encouragement they have given to me during the preparation of this book. In particular I would like to record my thanks to the late Professor R. Oliver, who originally suggested that the book should be written, to Dr S. M. Lewis and other former colleagues at Hammersmith Hospital and the Royal Postgraduate Medical School, to Mr R. C. T. Buchan, Dr D. H. Keeling and other colleagues at Plymouth General Hospital, and finally to Miss A. O'Connor for typing the manuscript and Mr G. B. Hodges for help in preparing the illustrations.

C. S. Bowring

Contents

Basic physics
of radionuclides

Atomic and nuclear structure

It is generally accepted that matter can be broken down into a limited number of substances known as the chemical elements and that the smallest unit of any element is the atom. Each atom may be regarded as consisting of a central nucleus, made up of positively charged particles called protons and uncharged particles called neutrons, surrounded by orbiting particles called electrons carrying negative electrical charges. The mass of a proton can be taken as equal to the mass of an neutron and is about 2000 times greater than the mass of an electron. The nucleus therefore contains almost the entire mass of the atom. In terms of size, however, the nucleus typically has a diameter 10^4 times smaller than that of the atom. (A typical atomic diameter would be 10^{-10}m.) The positive electrical charge on each proton is numerically equal to the negative charge on each electron and, as the number of electrons is equal to the number of protons, the atom as a whole is electrically neutral. Since the number of protons in the nucleus determines the number of electrons, and the number of electrons in turn determines the chemical properties of the atom, we find that the atoms of each element are characterized by having a certain specific number of protons in their nuclei. The number of neutrons in the nucleus, however, can vary, within limits, since they have no effect on the electronic structure of the atom. Typically, the lighter elements have approximately equal numbers of neutrons and protons in their nuclei, while in the heavier elements there is an excess in the number of neutrons over the number of protons. Not all combinations of neutrons and protons are stable and those which are not are termed radioactive.

Radioactive decay

When a nucleus does not have a stable combination of neutrons and protons it undergoes radioactive disintegration or decay. This involves the capture or expulsion of a charged particle by the nucleus and results in a change in the number of protons in the nucleus so that an atom of a different element is formed. The new nucleus is then usually left in the so-called 'excited state' with excess energy due to non-optimal arrangement of its neutrons and protons. In most cases the new nucleus immediately rearranges its internal structure so that it attains its ground energy state. In the process of this the nucleus emits its excess energy in the form of one or more photons of radiation known as gamma (γ) rays. Sometimes the new nucleus is already in its ground state and there is then no γ-ray emission. Sometimes, the excited state may be sufficiently stable for the nucleus to remain in it for a measurable length of time; in this case it is said to be an isomer and to be in a metastable state.

The original nucleus is often referred to as the parent, and the product of the decay process as the daughter. The radioactive decay process is shown diagrammatically in *Figure 1.1* The daughter may not be stable, of course, and would then undergo one or more further decay processes.

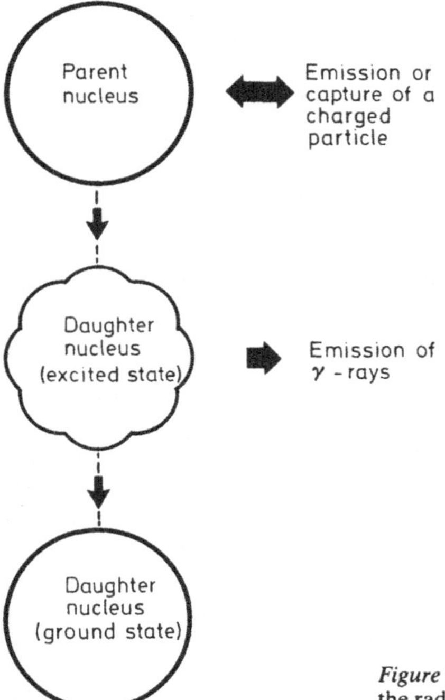

Figure 1.1 A diagrammatic representation of the radioactive decay process

Nuclides and isotopes

A nuclide is any specified nucleus with a stated atomic number, Z, (equal to the number of protons in the nucleus), a mass number, A, (equal to the number of protons plus neutrons in the nucleus), and in a defined nuclear state. It is normally assumed that the nucleus is in the ground state and the same symbolism is used as for atoms. Thus $^{51}_{24}Cr$ represents a nucleus of the element chromium with a mass number 51 and an atomic number 24. If, however, the nucleus is in a metastable state then the superscript 'm' is added so that $^{99m}_{43}Tc$ (technetium) would indicate the metastable state of $^{99}_{43}Tc$. In practice the atomic number is commonly omitted and the above nuclides are written ^{51}Cr, ^{99m}Tc and ^{99}Tc. An isotope is one of a group of nuclides having the same atomic number but different mass numbers (i.e. only differing from each other in the number of neutrons in the nucleus), and radionuclides and radioisotopes are radioactive nuclides and isotopes.

Nuclear radiation

The particle that is emitted or captured by the nucleus in the radioactive decay process, for all the radionuclides with which we shall be concerned, is either the beta (β) particle or the electron. The β-particle has a mass and charge numerically equal to that of an electron but may be either negatively or positively charged. The negatively charged particle is known as the β^- particle or simply as the β-particle. This has properties identical to an electron. The positively charged β-particle or β^+ particle is known as a positron.

The β-particles emitted in the decay of a particular radionuclide have various energies up to a maximum value which is characteristic of that radionuclide and usually have an average energy of approximately one-third of their maximum energy. In the case of decay by the emission of a β^- particle, γ-rays may or may not be emitted depending on whether or not the daughter is in the ground state, but in radionuclides that decay by the emission of a positron, two photons of energy 0.511 MeV (for the definition of this unit see below) are always emitted. This radiation, which is often called annihilation radiation, arises from the combination of the emitted positron when it has come to rest with an electron and the conversion of their joint masses into energy. The annihilation photons are emitted in opposite directions. An alternative mode of decay to positron emission is the capture by the nucleus of one of the innermost orbiting electrons. This results in the daughter atom having a hole in its electronic structure which is immediately filled by an electron from an outer orbit which causes the emission of the characteristic X-ray photons of the daughter atom.

Sometimes, instead of the γ-ray being emitted normally, its energy is given to one of the inner orbiting electrons which is then ejected from the atom with a kinetic energy equal to the difference between the energy of the γ-ray and its binding energy. When this happens the γ-ray is said to be internally converted and again, since the process results in a hole in the electronic structure of the atom, characteristic X-ray photons are emitted. It is perhaps worth noting that although X-rays and γ-rays of the same energy are indistinguishable, X-rays result from transitions between electron energy levels and γ-rays result from transitions within the nucleus of the atom.

In summary: radionuclides which decay by positron emission or electron capture always have γ-ray and/or X-ray photons associated with them; radionuclides which decay by β⁻ emission only have γ-rays associated with them if the decay is to an excited state of the daughter nucleus; and metastable radionuclides have only γ-rays associated with their isomeric transition to the ground state. Bremsstrahlung or braking radiation may, however, also be produced when high energy β-particles are absorbed, especially in materials of high atomic weight. Bremsstrahlung radiation is made up of X-ray photons the great majority of which have much less energy than that of the particles producing them.

The energy of all nuclear radiations is usually measured in electron-volts (eV), one eV being the energy gained by an electron when it moves through a potential difference of one volt. All β-particles are relatively easily absorbed by matter and usually only have a range of a few millimetres in tissue; γ-rays and X-rays of the same energy are, however, much more penetrating and are absorbed exponentially by a given material. Two different attenuation mechanisms are involved in the absorption of the energy of γ-rays and X-rays. First, there is photoelectric absorption where the entire energy of the X-ray or γ-ray photon is absorbed by an atomic electron and, secondly, there is 'Compton' scattering where the X-ray or γ-ray loses only a part of its energy to the electron and the photon continues with reduced energy in a different direction. Thus, while in the simplest case γ-rays all of the same energy may be emitted from the radionuclide, when these are detected many of them will have lost most of their energy. As a result, the detected photons have a range of energies similar to that shown in *Figure 1.2*.

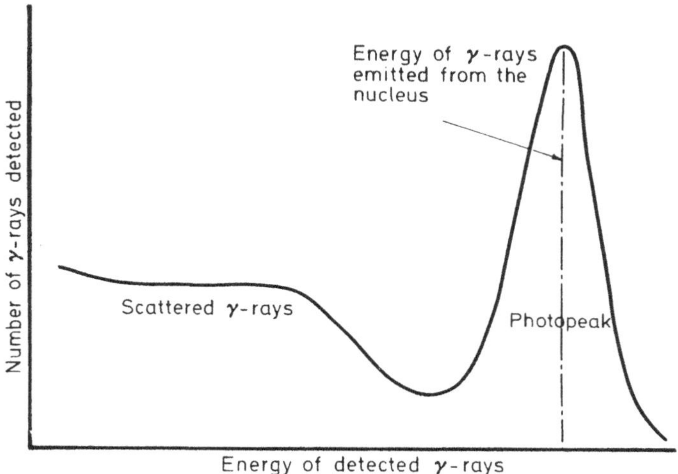

Figure 1.2 The observed energy spectrum for a source of ^{99m}Tc in tissue, which shows the unscattered γ-rays in the photopeak of the detector and the scattered γ-rays at lower energies

Since the β-particles are so easily absorbed, radionuclides that decay purely by the emission of β⁻particles with no associated γ-ray emission, are very difficult to detect outside the body unless any Bremsstrahlung radiation that the β-particles produce can be detected. Therefore, they are usually used only for *in vitro* work. Radionuclides that emit γ-rays can normally be detected easily outside the body, the optimal energy range for detection being 0.1–0.3 MeV, and these can be used for *in vivo* studies. Of the radionuclides that emit γ-rays the metastable radionuclides can often be administered in the largest quantities since, first, the amount of radiation energy absorbed by the patient is generally less owing to the absence of the easily absorbed β-particles and, secondly, because they mostly have relatively short radioactive half-lives.

Radioactive half-life

The rate at which the nuclei of a given radionuclide undergo decay is a characteristic property of that radionuclide and is unaffected by any physical or chemical condition. The greater the instability of the radionuclide the greater is the probability that it will decay in any time interval. This probability, therefore, is also characteristic of the radionuclide. It is not, however, possible to make any prediction about when any individual nucleus will actually decay. Fortunately, in practice, large numbers of atoms are almost always dealt with and, although the individual decay processes occur at random, the statistical behaviour of the group as a whole can be mathematically described with a high degree of accuracy.

The law of radioactive decay states that the average number of disintegrations occurring in one second in a sample of a radionuclide at any time is proportional to the number of nuclei present in the sample at that time.

Mathematically, the fraction of the parent nuclei remaining at time t is given by the expression in equation (1), where λ is called the radioactive decay constant and represents the fraction of atoms undergoing decay in unit time, $N(0)$ is the number originally present at time $t = 0$ and $N(t)$ is the number remaining after time t.

$$N(t)/N(0) = e^{-\lambda t} \tag{1}$$

The radioactive half-life is the time taken for the number of the parent nuclei to be reduced by one-half and this is related to the decay constant by the expression given in equation (2).

$$\lambda = 0.693/T_{1/2} \tag{2}$$

where $T_{1/2}$ is the radioactive half-life. Therefore, equation (1) can be rewritten in the form of equation (3) and it is this form which is normally used since the half-life is usually a more informative and easily remembered entity than the decay constant.

$$N(t)/N(0) = e^{-0.693t/T_{1/2}} \tag{3}$$

Thus if we consider the radionuclide ^{99m}Tc, which has a half-life of 6.0 hours, we can calculate the fraction remaining at any time t by calculating the quantity $e^{-0.693t/6.0}$, where t is measured in hours. This can be simply achieved on most calculators or by looking up tables of e^{-x} and *Figure 1.3* shows how this factor varies with time for ^{99m}Tc. Alternatively, the decay curve becomes a straight line if it is plotted on semilogarithmic graph paper as shown in *Figure 1.4*. In *Figure 1.4* the fraction remaining is plotted on the logarithmic scale and time on the linear scale. The line passes

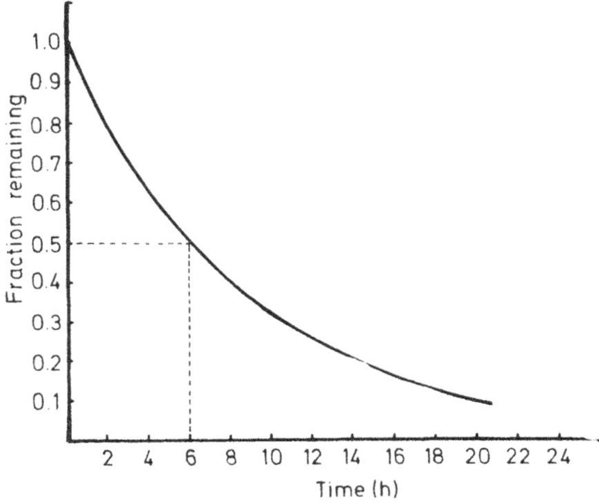

Figure 1.3 The radioactive decay of ^{99m}Tc plotted on linear graph paper

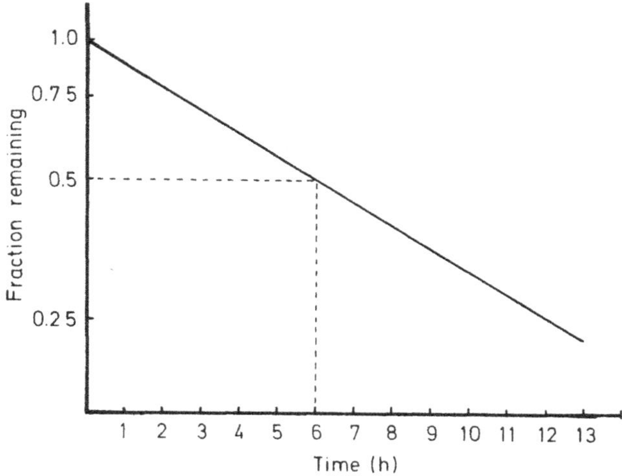

Figure 1.4 The radioactive decay of ^{99m}Tc plotted on semi-logarithmic graph paper

through the point at which half the parent nuclei have decayed at a time equal to the half-life, and many people prefer to plot out and use decay charts like this rather than to use equation (3).

As an example of the type of decay calculation that is often required, let us consider the calculation of the volume of ^{99m}Tc pertechnetate solution that is needed for labelling red blood cells for a spleen scan. For this investigation about 150 megabecquerel (MBq) of ^{99m}Tc is needed at the time of injection (for the definition of MBq see below). Let us assume that the proposed time of injection is 1130, that the time is now 1030, and that the stock solution contained 500 MBq of ^{99m}Tc/ml of solution at 0900. The volume required for labelling can now be calculated in the following way. At 1130 the activity of the stock solution will be reduced by the factor, 0.75, for 2.5 hours decay and thus by this time the stock solution will contain $500 \times 0.75 = 375$ MBq of ^{99m}Tc/ml. Therefore, the volume required for labelling is 150/375 $= 0.40$ml. (The actual quantity of radioactivity present at 1030 is given by $0.40 \times 500 \times 0.84 = 168$ MBq, where 0.84 is the decay factor for 1.5 hours.)

The unit of radioactivity

In the past the amount of radioactivity present in a sample has been specified in terms of a unit called the curie (Ci) which is, for

Table 1.1 Table for the conversion of the units used for measuring radioactivity

Becquerels	Curies
1 Bq	27 pCi
1 kBq	27 nCi
37 kBq	1 μCi
500 kBq	13.5 μCi
1 MBq	27 μCi
18.5 MBq	500 μCi
37 MBq	1 mCi
500 MBq	13.5 mCi
1 GBq	27 mCi
37 GBq	1 Ci

historical reasons, defined as that amount of radioactive material in which the number of disintegrations occurring per second is 3.7×10^{10}. However, the use of the curie is now being discontinued and the SI unit, the becquerel, should be used instead. One becquerel corresponds to one disintegration every second and hence there is a conversion factor of 3.7×10^{10} for converting curies to becquerels. All quantities of radioactive material mentioned in this book will be specified in becquerels and a conversion between the two units is given in *Table 1.1*. In practice, quantities of radioactive material in a range from 10^3 Bq to 5×10^8 Bq (from 1 kBq to 500 MBq) are normally dealt with.

The specific activity of a sample is a measure of the ratio of the quantity of the radionuclide in the sample to the mass of the same element in the sample and is normally given as the radioactivity per unit mass of the element or compound containing the radionuclide. In carrier-free radionuclides there are no stable isotopes of the radionuclide in the sample.

Statistical factors and measurement errors

Tracer work with radionuclides nearly always involves comparisons of the amount of radioactivity in various samples or organs of the body and these data are obtained by measuring the number of decay processes occurring in a given time from each sample. For these comparative measurements it is not necessary to determine the total number of disintegrations occurring in each sample, but it is necessary to ensure that the measurement conditions are constant for all the measurements. However, since radioactive decay is a random process, repeated measurements on the same sample may not give exactly the same result and thus it is important to be able to estimate the error in the result obtained.

The standard deviation of a single measurement of the number of disintegrations detected in a given time interval is equal to the square root of the total number detected, thus the larger the number detected the smaller the error becomes when expressed as a percentage of the result. In practice, if a count of N disintegrations is recorded then the true result has approximately a 68 per cent probability of lying between $N - \sqrt{N}$ and $N + \sqrt{N}$, a 95 per cent probability of lying between $N - 2\sqrt{N}$ and $N + 2\sqrt{N}$ and a 99.7 per cent probability of lying between $N - 3\sqrt{N}$ and $N + 3\sqrt{N}$.

It is usually found that it is necessary to correct any measurement for 'background', which is the number of counts recorded when there is no sample present. The actual result is then given by the difference between the sample count and the background count. In this case one standard deviation in the result is given by $\sqrt{(N + B)}$ where B is the background count (obtained over the same length of time as the sample count) and the same error is also present in the addition of the two counts. In the case of division or multiplication of one count by another the standard deviation of each count must be expressed as a percentage of the count (it is then known as the relative standard deviation). The relative standard deviation in the result is given by $\sqrt{(v_1^2 + v_2^2)}$ where v_1 and v_2 are the relative standard deviations of the two counts.

For example, if in a given counting time 10 000 counts are obtained from sample A, 2000 counts from sample B and 500 counts from the background, then, working to one standard deviation, the true count from sample A is likely to lie within the range $9500 \pm \sqrt{10\,500}$ ie 9500 ± 102 and the true count from sample B is likely to lie within the range $1500 \pm \sqrt{2500}$ ie 1500 ± 50. If the ratio of the quantities of radioactivity in these samples is now considered then the relative standard deviation for sample A is 1.08 $(100\sqrt{10\,500}/9500)$ and for sample B is 3.33 $(100\sqrt{2500}/1500)$, so that the relative standard deviation in the ratio is $\sqrt{(1.08^2 + 3.33^2)} = 3.50$. Thus one standard deviation in the value of the ratio of the radioactivity in sample A to that in sample B is $(9500/1500) \times (3.50/100) = 0.22$ so that the true ratio has a 68 per cent probability of lying within the range 6.33 ± 0.22.

Radionuclides in common use and radiation dosimetry

The radionuclides that are commonly used in haematology and which are mentioned or discussed in this book are listed in *Table 1.2* together with their mode of decay, radioactive half-life and principal emissions, and it can be seen that these all cover a large range.

The administration of any radionuclide to a patient involves the irradiation of that patient. The radiation dose received by the patient is a measure of the quantity of radiation energy absorbed in the patient's tissues and this depends on both the physical properties of the radionuclide (*see Table 1.2*) and on the way in which the radiopharmaceutical is metabolized. The SI unit used for the measurement of radiation dose is the gray and this is equivalent to the absorption of one joule of energy per kilogram of tissue (previously radiation dose was measured in rads; 1 rad≡0.01 gray). The unit used in practice in radiation protection and recommended for use by the International Commission on Radiological Protection is, however, the sievert, which is numerically equal to the gray for the radiations considered in this book. In *Table 1.3* typical figures are given for the approximate radiation dose that a normal adult would receive from each of the radionuclide tracers discussed in this book. The figure given is for

Table 1.2 The major emissions and properties of some of the radionuclides commonly used in haematology

Radionuclide	Decay mode	Half-Life	Maximum β-particle energy (MeV)	Principal photon energies (MeV)
^{3}H	β^-	12.3 years	0.02	–
^{32}P	β^-	14.3 days	1.71	–
^{51}Cr	EC	27.8 days	–	0.320
^{52}Fe	EC β^+	8.2 hours	0.80	0.511 0.165
^{55}Fe	EC	2.60 years	–	0.006
^{59}Fe	β^-	45.6 days	1.57	1.292 1.095
^{57}Co	EC	270 days	–	0.122 0.136
^{58}Co	EC β^+	71.3 days	0.47	0.810 0.511
^{99m}Tc	IT	6.05 hours	–	0.140
^{111}In	EC	2.81 days	–	0.173 0.247
^{113m}In	IT	99.8 minutes	–	0.393
^{125}I	EC	60.2 days	–	0.035 0.028
^{131}I	β^-	8.05 days	0.806	0.364

Key to decay modes: β^- decay by emission of a β^- particle, EC decay by capture of an orbiting electron, β^+ decay by emission of a positron, IT, isomeric transition.

Table 1.3 The radiation doses to adults from the radionuclide tracers used in haematology

Tracer	Critical organ	Dose (mSv/MBq)
^{3}H-DFP	Blood	1
^{32}P-DFP	Blood	8
^{51}Cr-red cells	Spleen	1
^{51}Cr-heat damaged red cells	Spleen	20
^{51}Cr-platelets	Spleen	1
^{52}Fe-transferrin	Spleen	2
^{55}Fe-transferrin	Spleen	7
^{59}Fe-transferrin	Spleen	34
^{57}Co-vitamin B_{12}	Liver	40
^{58}Co-vitamin B_{12}	Liver	70
^{99m}Tc-red cells	Blood	0.02
^{99m}Tc-heat damaged red cells	Spleen	0.3
^{111}In-platelets	Spleen	6
^{111}In-transferrin	Whole body	0.2
^{111}In-white cells	Spleen	2
^{113m}In-transferrin	Blood	0.02
^{113m}In-colloid	Liver/spleen	0.02
^{125}I-HSA	Blood	1
	Thyroid	200
^{131}I-HSA	Blood	2
	Thyroid	80

the body organ which receives the highest radiation dose (sometimes known as the critical organ) and is given in mSv/MBq of radioactivity administered. To put these figures in perspective, the natural background radiation to which the whole body is exposed in a year varies from 0.7 mSv to about 1.5 mSv in different parts of the United Kingdom and is as high as 10 mSv in some parts of the world. It must be stressed that all the figures in *Table 1.3* are approximate and assume normal metabolism and may therefore be grossly inaccurate in individual patients. The dose to the thyroid when iodinated HSA is used illustrates this. As can be seen, two figures have been given. The thyroid dose is so high because of the release of the radioactive iodine into the exchangeable-iodine pool of the patient when the albumin is broken down. Subsequently, the thyroid treats the radioactive iodine in the same way as normal dietary iodine and a large amount is taken up into the gland. However, this high radiation dose to the thyroid can be avoided by administering a thyroid blocking agent to the patient (as is discussed in Chapter 3) and when this is done it is the blood which becomes the critical organ. Also worthy of note are the relatively low doses that are obtained with the metastable radionuclides, and which arise primarily from the short half-lives and lack of emission of β-particles.

It should perhaps be mentioned at this stage that only medical practitioners specially certified under 'The Medicines (Administration of Radioactive Substances) Regulations 1978' or those working under their direction are legally allowed to administer radioactive materials to patients in the United Kingdom. Before commencing any work using radionuclides, the local department of Nuclear Medicine or Medical Physics should be contacted with a view to obtaining practical instruction on safe working procedures.

References

The following textbooks have useful sections that can be recommended for further reading.

BELCHER, E. H. and VETTER, H. (1971) *Radioisotopes in Medical Diagnosis.* London: Butterworths
OLIVER, R. (1971) *Principles of the Uses of Radioisotopes in Clinical and Research Investigations.* Oxford: Pergamon
PARKER, R. P., SMITH, P. H. S. and TAYLOR, D. M. (1978) *Basic Science of Nuclear Medicine.* Edinburgh: Churchill Livingstone

2

Instrumentation

The instrumentation needed for the detection and measurement of the radionuclides used diagnostically in haemotology is required to carry out five functions. Namely, the measurement of relatively large quantities of radioactivity used for imaging; the measurement of radioactivity in samples taken from the patient; the measurement of radioactivity in the whole body or organs in the body; the visualization of the distribution of the radionuclide in the patient and the checking and monitoring of the working environment. The instrumentation comprises the following:

1. Ionization chambers;
2. Sample counters;
3. Uptake and whole-body counters;
4. Radionuclide imaging equipment;
5. Radiation detectors (discussed under sample counters).

In this chapter the basic principles of these items of equipment are discussed and further information will be found in the succeeding chapters.

Ionization chambers

These instruments are not normally found in the haematology department but in the departments of nuclear medicine or medical physics, which usually operate as the central supplier of radionuclides in the hospital. A diagram of a simple ionization chamber is shown in *Figure 2.1* and the principle of its operation is very simple. The radiation emitted from a radionuclide placed in the centre of the chamber ionizes the gas between the two electrodes of the chamber, and the ions that are produced move to the electrodes under the influence of the electric field applied between them. This results in a small, but measurable, electric current flowing through the circuit. The magnitude of this electric current varies in direct proportion to the quantity of the radionuclide present. A potential difference of 100–200 volts is usually applied between the electrodes. The sensitivity of the chamber varies with each radionuclide and may also change with differences in the volume of the radionuclide. It is therefore necessary for the chamber to be calibrated with standard sources of each radionuclide. In the newer chambers there are usually push-button settings

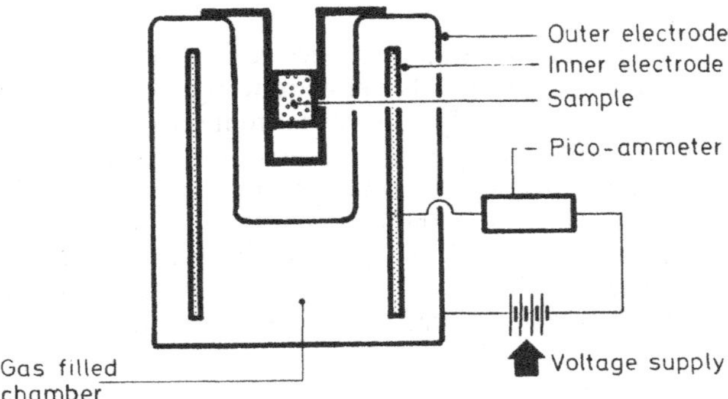

Figure 2.1 A cross-section through an ionization chamber of the type used for assaying radionuclides

for each of the commonly used radionuclides so that a direct reading can be obtained of the quantity of radioactivity present. Some chambers, instead of being filled with air at atmospheric pressure, are filled with another suitable gas (usually argon) under pressures of up to about 20 atmospheres, which increases the chamber's sensitivity. For most radionuclides the useful measurement range of the ionization chamber is from about 100 kBq upwards.

Sample counters

Whereas the ionization chamber measures the overall ionization current produced in the instrument, the sample counters detect and count the individual γ-rays, X-rays or β-rays that are emitted by the sample. These instruments are capable of detecting small quantities of radioactivity and are normally used over the range 1 Bq–100 kBq, and small handheld versions of these instruments are used as contamination monitors. Three types of detection system are commonly used: the Geiger–Müller tube counter (GM counter), the scintillation counter and the liquid scintillation counter—the use of the first and last of these being effectively limited to the detection of β-particles. In all cases the sensitivity varies with the geometry of the sample being counted so that it is essential to ensure that all samples in a batch have the same form.

The Geiger–Müller counter

The basic GM counter consists of two electrodes, an outer cylinder which forms the negative cathode, and a very thin wire which runs along the axis of the cylinder and forms the positive anode, and the whole is sealed and filled with a suitable gas mixture (see below). A typical tube is shown in *Figure 2.2*. The radiation passes through

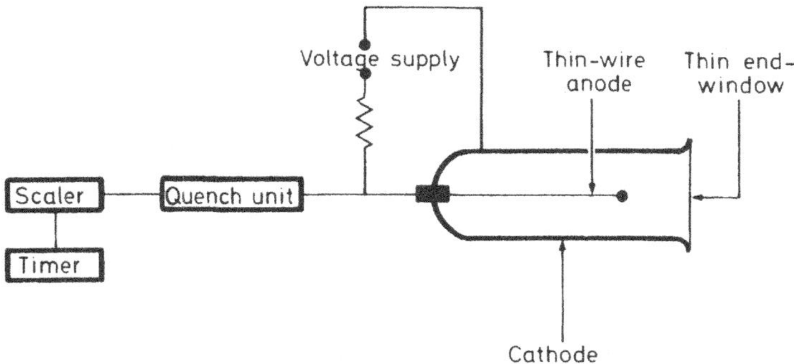

Figure 2.2 An end-window Geiger counter and its electronics

the detector and ionizes the gas in the tube and the voltage and gas pressure are then adjusted so that the ionization produced is increased by gas multiplication. This occurs when the electrons released in the primary ionization process can gain enough energy to produce further ionization. The result of this in a GM tube is that from one single ionizing event an avalanche of up to about 10^{10} electrons strikes the anode, instead of the one electron that would register the current in an ionization chamber. The gas that fills the counter is normally argon to which a trace of bromine vapour (known as the quenching agent) is added to prevent the counter from refiring spontaneously. Each β-particle or X-ray or γ-ray photon detected therefore leads to an electrical pulse which can be counted with a simple scaler. The quench unit (*see Figure 2.2*), as opposed to the quenching agent, ensures that the detector remains insensitive for a fixed time (the dead time) after each event. The scaler is generally operated in conjunction with a timer so that either the number of events recorded in a fixed time is measured or, alternatively, the time taken to record a fixed number of counts is measured. Since the dead time of a GM detector is typically about 300 microseconds it is often necessary to make a dead-time correction to the recorded count. Thus, if N counts are recorded in a time of T seconds, the actual time for which the counter has been actively operating is $T-Nt$, where t is the dead time. Therefore, the true count that should have been recorded in time T is not N but $TN/(T-Nt)$ or $N/(1-Nt/T)$.

The correct operating voltage for any particular GM counter is determined by plotting the count rate obtained from a source of radioactivity as a function of the voltage applied between the electrodes. The characteristic curve obtained should be similar to that shown in *Figure 2.3*. At low voltages the gas amplification is

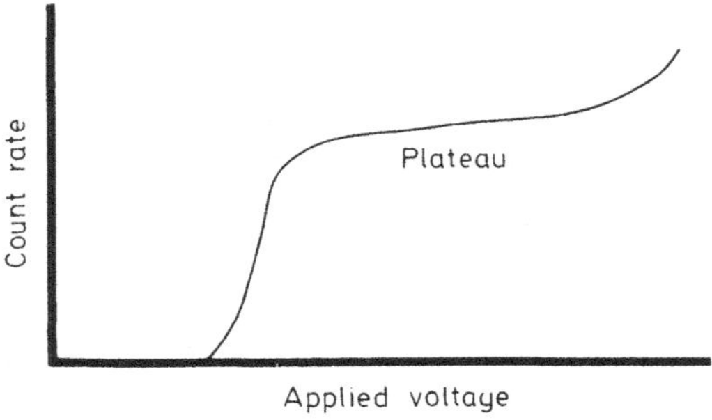

Figure 2.3 The characteristic curve of a typical GM tube

insufficient for the device to function and no pulses are observed, but at a particular voltage operation starts and the observed count rate rises rapidly and then stabilizes, so that the response remains reasonably constant over the range of voltage known as the plateau. Above the plateau voltages the count rate again rises rapidly and sparking or continuous discharge occurs between the electrodes. It is not critical at what voltage the tube is operated as long as it is on the plateau but, in practice, a point towards the lower voltage end is normally chosen. In a good tube the plateau is long and nearly flat, whereas a tube at the end of its useful life will probably have a very short plateau region or even none at all. It is necessary, therefore, to check the performance of GM tubes at intervals by plotting their characteristic curves. Typical operating voltages are in the range 400–600 volts for tubes containing bromine (halogen) quenching gas and 1200–1500 volts for older type tubes using alcohol vapour as the quenching agent.

In sample-counting applications these detectors are used only for counting β-particles since scintillation counters (*see* below) are much more sensitive to X-rays and γ-rays. However, GM counters are commonly used for monitoring purposes and two small portable bench monitors are shown in *Figure 2.4*.

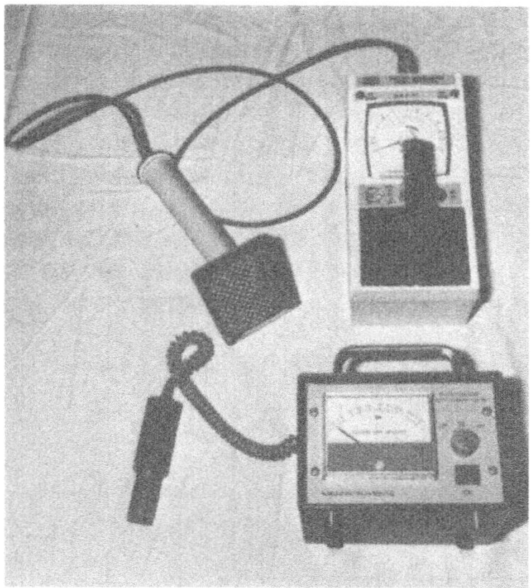

Figure 2.4 Two small portable contamination monitors

As stated previously, β-particles are absorbed very easily and it is therefore necessary to have a 'thin window' in the detector tube through which the β-particles can enter the detector. Once inside the detector a single β-particle is almost certain to produce some ionization and therefore be detected. Most of the GM tubes used in sample counters are made in the bell form shown in *Figure 2.2* and have a thin (and very fragile) mica window at their base. The diameter of the window is not normally more than about 30 mm. However, even the thin mica absorbs some of the β-particles and thus reduces the detector's sensitivity, and in the case of very low-energy particles such as those emitted in the decay of ^{3}H any window provides too much absorption for detection to be efficient. There are two ways round this problem. The first, which is not very commonly used, is to burn the sample to form a radioactive gas which is then used to fill the tube. The second is to place the entire sample inside the tube. In this latter method the counter has to be demountable and a suitable gas mixture is flushed through the detector during operation. Such gas flow counters have increased sensitivity, but the technique of liquid scintillation counting (*see* below) is now more usually adopted.

The scintillation counter

The scintillation counter is, as the name implies, a detection system which counts scintillations. When radiation is absorbed by some materials, the material scintillates and the amount of light given out is proportional to the energy of the radiation absorbed. The detector material is normally a specially-grown single crystal of sodium iodide to which small amounts of thallium have been added [NaI (Tl)], although other crystals such as cesium iodide (CsI) are used sometimes and very large detectors are often made from special plastics. A typical detector assembly is shown in *Figure 2.5*. Since the crystal is hygroscopic it has to be protected

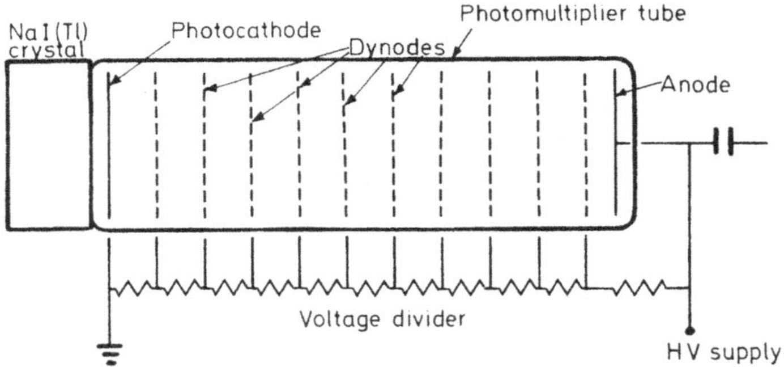

Figure 2.5 A scintillation counter detector assembly comprising detector crystal, photomultiplier and resistor chain which is used to power the dynodes of the photomultiplier

from atmospheric moisture and is normally hermetically sealed in an aluminium can. Unfortunately this has the effect of absorbing β-particles so that scintillation counters are generally suitable for use with X-rays and γ-rays only.

A photomultiplier (PM) tube is positioned above the detecting crystal to measure the amount of light produced. The input face of this has to be in very good optical contact with the crystal and the whole assembly must be completely light tight. Light photons entering the PM tube from the crystal strike a photosensitive surface (the photocathode) and cause it to emit electrons. These electrons are accelerated in the vacuum of the tube by the potential difference of about 100 volts applied between the photocathode and the first tube electrode (or dynode) and there produce further electron emission. These electrons are then accelerated to the next dynode where yet more electrons are emitted. This process is repeated through, typically, about 12 stages so that, at the final electrode (the anode), an easily detectable electrical pulse is produced.

As has already been noted, the amount of light entering the PM tube is proportional to the energy absorbed in the crystal and, therefore, the size of the output electrical pulse also depends on this. Thus, whereas in GM detectors each detected event produces a signal of approximately the same size, in scintillation counters the size of the signal varies with the energy of the detected radiation. It is therefore possible with scintillation counters to discriminate against low-energy radiation and count only the unscattered γ-rays in the photopeak (*see Figure 1.2*). The electronic device used for this is known as a pulse height analyser (PHA) and usually has two controls which are either an upper and lower energy level or a level and window width and these are used to set a 'window', or counting channel, so that only events producing a signal pulse with an amplitude falling within this range may be transmitted to the scaler for counting.

For all scintillation counting equipment it is advisable to plot the spectra of the commonly counted radionuclides at intervals in order to check the PHA settings. This is done by plotting the variation in count rate as the window position is altered: a spectrum similar in form to that shown in *Figure 1.2* should be obtained. If there is any change in the position of the photopeak since it was last checked, then the window settings used for counting should be changed, or the high voltage applied to the PM tube should be adjusted or the amplifier gain should be changed in order to bring the photopeak back to its previous position.

A block diagram of a scintillation counting system is shown in *Figure 2.6*. In the case of the GM tube counting system (*see Figure 2.2*) the signal pulses are of sufficient size to operate the scaler without further amplification, but in a scintillation counter system they generally do need further amplification. The operation of the PHA has already been discussed and the scaler and timer are operated in the same way as with a GM tube system. The ratemeter is entirely optional in sample counters and merely gives a visual indication of the count rate. If it is present it is generally used for rapidly setting-up the PHA window: the greatest count rate is normally obtained with the window centred over the photopeak. In contamination monitoring instruments the rate-meter is commonly the only output display. Most modern commercial scintillation sample counters have up to three separate

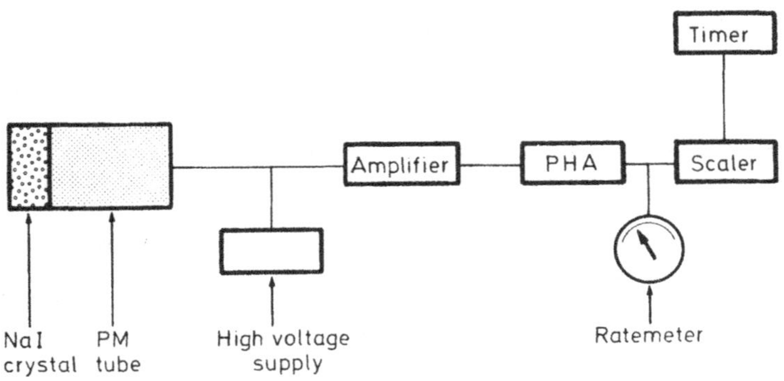

Figure 2.6 A typical scintillation detector with its associated electronics

PHAs and scalers, so that when samples contain more than one radionuclide the quantity of each can be measured simultaneously. In addition, they usually incorporate an automatic print-out of the results and automatic changing of the samples so that large batches of samples can be counted with ease.

When sample counting, a 'well-crystal' is usually used as the detector. This is a crystal with a well cut into it into which the sample is placed. This increases the sensitivity of the detector because the crystal then surrounds the sample. The detectors used typically have diameters of about 125 mm and wells of diameters up to about 25 mm that take sample tubes containing up to 20 ml of fluid. Systems for counting larger samples such as 24-hour urine collections, faecal collections or human extremities, such as hands or forearms, generally comprise either two larger crystals in between which the sample for counting is placed or a large well made of plastic scintillator. The use of plastic scintillators rather than NaI for large detectors arises primarily from the enormous increase in cost of NaI detectors due to the difficulties of production of large crystals. Plastic scintillators are not used in smaller detectors, however, because they have a much lower sensitivity to γ-rays than NaI and also a poorer energy response. The dead-time of scintillation detectors is typically about one-hundredth of that of GM detectors so that it is seldom necessary to make dead-time corrections.

Liquid scintillation counting

As well as using solid scintillating materials it is possible to use special solutions containing liquid scintillators. The radioactive material to be counted is suspended or dissolved directly in this liquid, which is usually based on an organic solvent, and the whole sample then takes the place of the detector crystal so that even very low energy radiation—whether β-particles or X-ray or γ-ray photons—can be detected. The technique is most commonly used for samples containing the radionuclides ^3H and ^{14}C and, in haematology, ^{55}Fe.

Although liquid scintillation counting overcomes the problem of getting the β-particles into the scintillator, there are other problems. Because the energy of the β-particles is usually very low, the size of the signal pulses produced is very small and is often comparable with the random electrical 'noise' pulses which are generated in the PM tube. This results in very high background counts which effectively reduce the accuracy of the measurement (*see* Chapter 1). The problem can be reduced in two ways. First, the temperature of the PM tube can be reduced by refrigerating the detector system and samples to reduce the noise level. Secondly, a coincidence counting method can be used. If two PM tubes are used to view the sample, a scintillation in the sample should be observed simultaneously in both tubes. Thus, if coincident pulses from both PM tubes only are counted, most of the background noise pulses can be rejected.

Another practical problem associated with liquid scintillation counting is the reduction in efficiency due to chemical and colour quenching. Chemical quenching is caused by the chemical absorption of the energy released during the decay process before it can be transferred to the scintillator and colour quenching is caused by the absorption of the emitted light photons in the solution before they can reach the PM tubes. The effect of quenching is to change the energy spectrum of the detected β-particles as shown in *Figure 2.7*, by reducing both their apparent energy and their number. It is, therefore, very important that great care is taken in sample preparation.

Three main methods are used to correct for the quenching effect and facilities for making these corrections are normally incorporated in commercial equipment. The primary method involves counting each sample twice, once normally and then again after the addition of a small known quantity of the radionuclide being counted. From the increase in the measured counts the counting efficiency can be calculated and used to correct the original count. The second method involves dividing the measured spectrum into two counting channels covering different parts of the energy

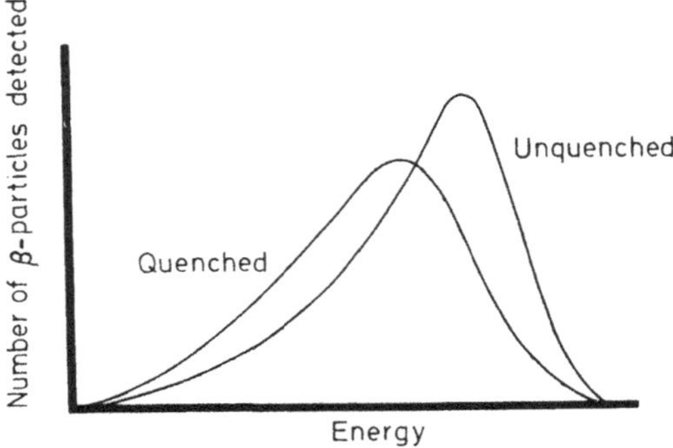

Figure 2.7 The energy spectra of unquenched and quenched scintillations in a liquid scintillation counting system

spectrum. Because quenching alters the ratio between the number of counts detected in each channel, measurement of this ratio provides a correction factor, provided calibration curves have been previously prepared using the primary method. The third method utilizes a γ-ray source which is positioned close to the sample. The γ-rays from this source interact with the liquid scintillator and the increase in the number of counts recorded shows the relative amounts of quenching present in each sample. As with the second method, this also has to be calibrated at some stage using the primary method. In practice, once the calibration charts have been prepared, either the second or third methods are generally used.

Uptake and whole body counters

Uptake counters, external counters or surface counters, as they are variously known, are used to obtain a measure of the uptake or loss of a radionuclide in or from an organ of the body. Usually, several measurements are made over a period of time so that the variation in uptake with time can be recorded. Whole body counters on the other hand are used to measure the total quantity of the radionuclide in the whole body. Both types of system generally employ NaI scintillation detectors.

Uptake counters

Since these counters are used in an attempt to measure the quantity of radioactivity in one specific organ only, it is normally necessary to restrict the field of view of the detector so that, in the ideal case, radiation from the organ of interest only is detected. The detector is usually placed on the skin above the organ of interest and the field of view is limited by using a collimator between the detector crystal and the skin. Usually the collimator is made of lead and in the simplest case consists of a single cylindrical hole in a block of lead. Such a collimator is shown in *Figure 2.8*: it

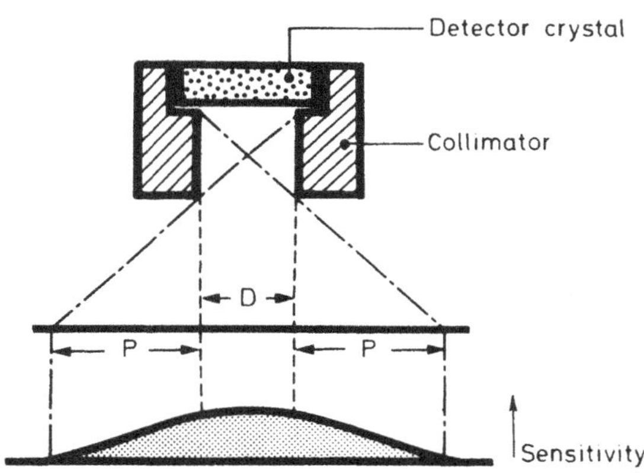

Figure 2.8 A simple collimator and its field of view

illustrates several important points in the design of uptake counters and imaging equipment. Ideally, the field of view is restricted to that immediately below the aperture of the collimator (D in the diagram) but an additional area around this called the penumbra (P in the diagram) is also detected with reduced sensitivity. Thus, if a small point source of radioactivity is moved across the entire field of view, the sensitivity as a function of position will be similar to that shown at the bottom of *Figure 2.8*. Obviously, the penumbra should be kept as small as is practical. Its size can be reduced by increasing the length of the collimator, although it can never be removed entirely, and it can be seen that the size of the penumbra will always increase as the distance from the face of the collimator increases. In general, the sensitivity of the collimator is closely related to the area of its field of view but two particular situations need to be considered. First, if the organ is smaller than the field of view, the sensitivity decreases as the separation of the detector from the organ increases, because the solid angle subtended by the organ at the detector decreases. Secondly, if the organ is larger than the field of view, then, as the separation increases, more of the organ comes into the field of view with the result that the solid angle remains constant and the sensitivity is maintained. In practice, of course, there is also a decrease in the sensitivity as separation increases, due to the absorption of radiation by the tissue and by the air that lie between the organ and the detector.

The electronics required for an uptake counter are essentially identical to those required for scintillation sample counters.

Whole body counters

In metabolic tests that use radionuclides it is often desirable to measure the retention of the radionuclide in the whole body over a period of days or weeks. Although this can be done indirectly by measuring the total amount of the radionuclide excreted, it is usually easier and more accurate to measure the amount retained directly. Since the radionuclide is likely to change its distribution in the body over the course of the investigation, it is essential that the sensitivity of the counter is independent of the location of the radionuclide in the body. Two systems are commonly used; shielded room systems and shadow shield systems, and both these types of systems may also be used for counting large-volume samples if no large sample counter is available.

Shielded rooms usually consist of a specially built low-background room with thick steel, lead or, occasionally, chalk walls. The arrangement of detectors in the room and the amount of built-in shielding depends basically on the amount of finance available and the sensitivity required. The simplest systems consist of one large detector crystal placed two metres or more from the couch on which the patient lies. This ensures that the distance of the detector from each part of the body is approximately the same. The more sophisticated systems may have 12 or more crystals positioned around the patient (generally not more than a few centimetres away) or large volumes of plastic scintillator, or tanks of liquid scintillator, which surround the patient.

The principle of the shadow shield system is shown in *Figure 2.9*. Usually two crystal detectors are used, one above and one below the patient, and the shielding around them is arranged so that there is no direct 'line of sight' into either of the detector crystals from outside the shielding. In order to measure the total body radioactivity the patient is slowly moved through the field of view of the detectors; the total count recorded is proportional to the whole body radioactivity. These shadow shield systems are probably the most useful type of whole body counter, although less sensitive than the more sophisticated shielded-room systems, because they are much cheaper and are also sufficiently sensitive to be able to detect whole body activities down to about 1 kBq in counting times of about 20 minutes.

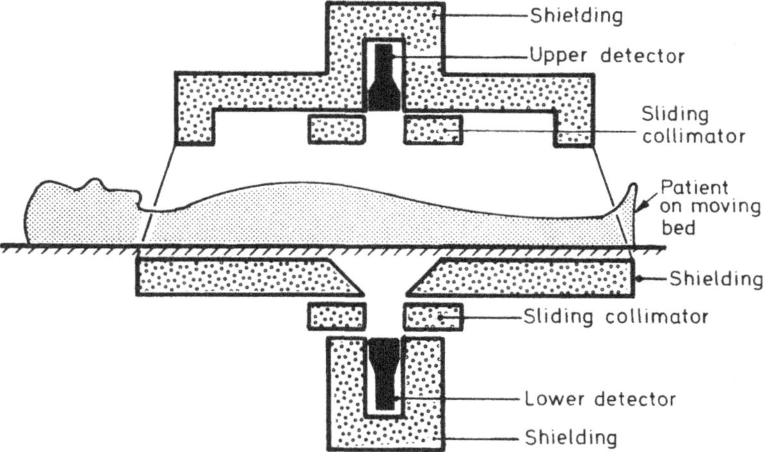

Figure 2.9 A section through a shadow shield whole-body counter

In addition, it is possible to use shadow shield systems to obtain a profile of the distribution of the radionuclide in the patient by recording the variation in count rate as the patient is moved through the field of view of the detector.

Radionuclide imaging equipment

Two instruments are commonly used to provide images of the distribution of a radionuclide within a patient. These are the rectilinear scanner and the gamma camera, although some hybrid systems and tomographic scanning systems are also available.

The rectilinear scanner

In the simplest scanners a single scintillation detector is moved in a rectilinear raster pattern over the patient and the recorded count rate is plotted on paper or film as a series of dots, their colour or density varying with the count rate at that position. The position of the plotter is synchronized to the position of the detector by a mechanical or electrical linkage. The basic detection system is similar to that of an uptake counter except that the movement and positioning of the detector is motorized and the collimation is improved in order that the field of view at any one time may cover only a very small part of the organ to be imaged. As already mentioned, the display is made up of a series of dots and the number of dots in any part of the image is proportional to the number of γ-rays detected from that part of the organ. Focused collimators made up of many tapered holes (*see Figure 2.10*) are

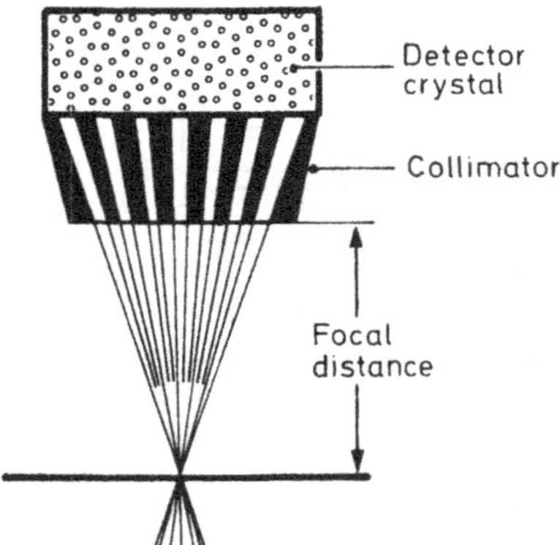

Figure 2.10 A focused collimator, typical of those used in scanning equipment, showing its field of view. (The penumbra is not shown)

used to reduce the size of the field of view and to produce a small volume with maximum sensitivity at a fixed distance (the focal distance) below the collimator face. As with the simple collimators used in uptake counters, good sensitivity and good positional resolution require opposing design considerations and a compromise is always required.

In haematology it is often of more interest to obtain an exact measurement of the quantity of radioactivity in an organ rather than just to visualize it, and for this it is necessary to have a scanner with two detector heads. The detectors are positioned opposite each other, one above and one below the patient, and are fitted with specially designed long focal length collimators so that the sum of the sensitivities of the two detectors through the patient may be approximately constant. The total measured count rate from an organ can then be made to be independent of the depth of the organ in the patient so that the number of counts recorded in the image is proportional to the amount of radioactivity present. A dual-detector scanning system with the patient's couch removed is shown in *Figure 2.11*.

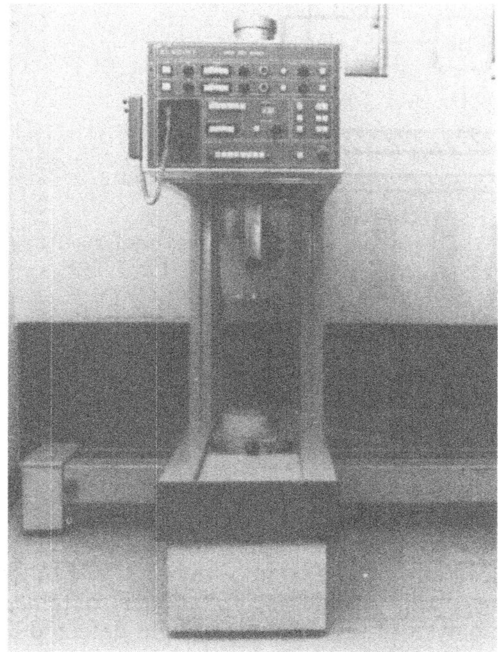

Figure 2.11 A dual-detector rectilinear scanner. The detector heads move laterally along the two arms of the yoke and the yoke moves along the length of the track. The patient's couch (not shown) fits between the two detectors

The gamma camera

The gamma or scintillation camera utilizes a larger NaI detector crystal (typically about 400 mm in diameter) which simultaneously views all parts of the organ to be imaged. Images can usually be obtained faster than with a scanner, and, since the whole organ is imaged at once, the instrument can be used to investigate fast dynamic processes. A diagram of a typical gamma camera system is shown in *Figure 2.12*. The collimator in front of the detector crystal usually consists of a great many small parallel holes in order

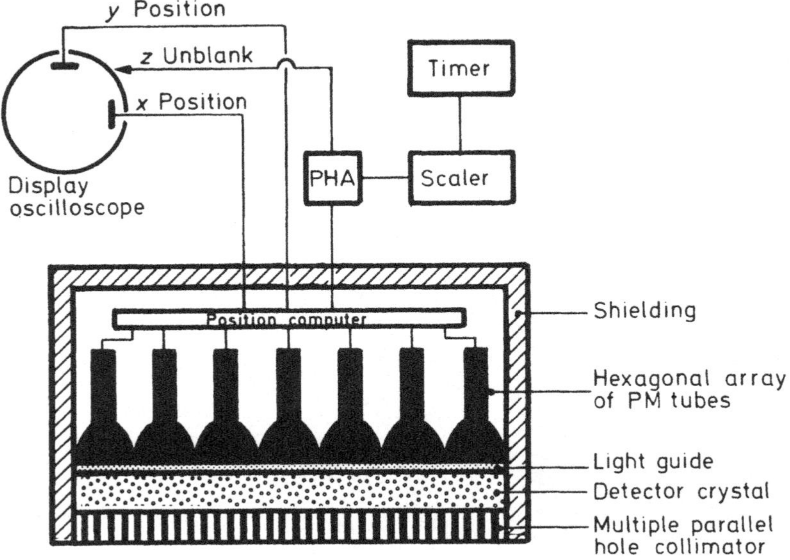

Figure 2.12 A typical gamma camera and its associated electronics

that the effective field of view may be the full size of the detector crystal. The crystal is normally viewed by an array of 37 PM tubes and the position in the crystal of each scintillation is obtained electronically from an analysis of the amounts of light detected in each tube. The positional signals are fed to a display oscilloscope and, if the total signal strength (ie the sum of the signals from all the PM tubes) falls within the window set on the PHA, a bright flash is momentarily displayed on the oscilloscope in the detected position. The final image is built-up from a time exposure of the oscilloscope display. A typical gamma camera is shown in *Figure 2.13*.

Unfortunately, the sensitivity of the gamma camera falls off rapidly with increasing distance from the collimator face with the result that even if two camera detection heads are used, one above and one below the patient, and their outputs are summed, there is still a large variation in sensitivity through the patient. Because of this, the dual head rectilinear scanner is generally preferred for quantitative imaging although better quality images are usually obtained with a gamma camera. It is, nevertheless, possible to obtain quantitative images with a gamma camera.

Figure 2.13 A gamma camera and its control console

Hybrid systems

There are several hybrid imaging systems now available. These utilize a long thin crystal as the detector. The detector head is positioned across the patient and then scanned along the length of the patient. The lateral positional information is obtained on the gamma camera principle and the longitudinal information is obtained from the detector motion. These systems can be configured to give quantitative information by the use of suitable collimation and two detectors. They may be expected to become more common over the next few years as the older rectilinear scanners are replaced.

Tomographic scanning systems

Tomographic scanning systems are imaging devices that are used to obtain cross-sectional images of the distribution of a radionuclide in the body, and operate in a very similar way to the well-publicized X-ray tomographic scanners. Information is recorded in a computer from a series of scans or images recorded at different angles, and the computer is then used to construct cross-sections of the distribution of the radionuclide. These images are essentially quantitative, but in order to measure the quantity of radioactivity in an entire organ accurately it is usually necessary to combine several images. At present, these devices are very expensive and it remains to be seen whether they will find a place in routine radionuclide imaging.

References

The following text books can be recommended for further reading.

HINE, G. J. (1967) *Instrumentation in Nuclear Medicine*, Vol. 1. New York: Academic Press

HINE, G. J. and SORENSON, J. A. (1974) *Instrumentation in Nuclear Medicine*, Vol. 2. New York: Academic Press

ROLLO, F. D. (1977) *Nuclear Medicine Physics, Instrumentation and Agents*. Saint Louis: C. V. Mosby Company

3

The measurement of red cell volume and plasma volume

One of the simplest diagnostic tests that uses radionuclides is the measurement of blood volume, and this is achieved by labelling the two major components of the blood, the red cells and the plasma, with known quantities of different radionuclides and then measuring their dilution. Although the haemoglobin level, the red-cell count and the packed cell volume (PCV) often reflect the total red-cell volume, they do not invariably do so, and the use of their measurement alone can often be misleading. This is particularly the case if the plasma volume is markedly larger or smaller than normal, because increased plasma volume can produce an apparent anaemia when none in fact exists and similarly a decreased plasma volume can mask a genuine anaemia. The measurement of total blood volume and that of its principal component parts, the red cell volume and the plasma volume, therefore has an established place in diagnostic haematology. In practice the measurement has been most used in the diagnosis of polycythaemia, where it is necessary to demonstrate an absolute increase in the red cell volume in order to confirm the diagnosis, and in the investigation of sudden haemorrhage and blood loss of unknown severity.

When measuring blood volume it is normal to measure the red cell volume and the plasma volume independently and simultaneously, rather than to measure the red cell volume or the plasma volume alone and then estimate the other using a measurement of the PCV. It is often stated that the ratio of the whole-body PCV to the venous PCV is 0.91 and this figure has been used for such estimations. This procedure cannot, however, be recommended (*see* below).

The principle of the technique used to measure blood volume is, as already mentioned, that of dilution analysis. A small known quantity of a tracer is injected into the blood circulation and, after allowing time for complete mixing, blood samples are taken. The concentration of the tracer in the blood samples is then measured, and from this the volume into which the tracer has been diluted can be simply calculated. In the past, non-radioactive tracers were used for these measurements; for example Evans' blue dye for marking plasma, and donor red cells of a different MNSs blood group for marking the red cell population. However, for routine measurements these methods have now been completely replaced by the use of radioactive tracers.

The most commonly used radioactive labels for red cells are ^{51}Cr in the form of sodium chromate and ^{99m}Tc in the form of sodium pertechnetate, although other labels are used occasionally. The tracers most commonly used for the measurement of plasma volume are ^{125}I or ^{131}I labelled human serum albumin (HSA), and ^{113m}In labelled transferrin, although ^{59}Fe and ^{52}Fe can also be used. The relative merits of each of these tracers are discussed below.

Red cell labels

The most accurate measurement of red cell volume can normally be made with ^{51}Cr and the use of this label has several advantages. First, the labelling procedure is straightforward, secondly the amount of elution (loss of the label from the labelled cells) is negligible over the time required for the measurement, and thirdly its γ-ray emission can be distinguished easily from that of ^{125}I and satisfactorily from that of ^{113m}In and, therefore, the plasma volume can be measured simultaneously. The primary disadvantage of ^{51}Cr for volume measurements is that its good labelling properties and its relatively long radioactive half-life of 28 days mean that the blood radioactivity remains high for some time, making repeat measurements difficult. The radionuclide ^{99m}Tc, on the other hand, has a much shorter half-life of 6 hours and thus measurements can be repeated easily, but unfortunately there is a spontaneous elution of the label *in vivo* of the order of 4–10 per cent/hour (Korubin, Maisey and MacIntyre, 1972; Ferrant, Lewis and Szur, 1974; Jones and Mollison, 1978) and the labelling procedure is more involved than that using ^{51}Cr. ^{32}P as sodium phosphate was formerly a popular alternative to ^{51}Cr (Mollison, Robinson and Hunter, 1958), but is not now commonly used because it is eluted from the cells at a rate of about 10 per cent/hour and also because ^{32}P is a pure β-emitter and sample preparation and counting is more involved than for γ-emitters. A short-lived cyclotron radionuclide that has been used is ^{11}C as carbon monoxide (Glass *et al.*, 1968) but, because of its very short half-life, it cannot be used at hospitals that do not have their own cyclotron and, even in these, it is not in common use.

The choice of red cell label thus lies between ^{51}Cr and ^{99m}Tc and the decision about which to use in a particular situation will probably depend on the other tests planned for the patient, although, if the mixing time is likely to be prolonged, it may be preferable to use ^{51}Cr because of its lack of elution. For instance, if a study of red cell survival is required ^{51}Cr would probably be used because the labelled cells could then also be used for the survival measurements (*see* Chapter 4), but, if the size of the splenic red cell pool is to be determined, then the total red cell volume could be determined at the same time using the same ^{99m}Tc-labelled cells (*see* Chapter 7). Alternatively ^{99m}Tc might be used if it was known that the measurement would need to be repeated in the near future. Details of methods used to label red cells with ^{51}Cr or ^{99m}Tc are given in Appendix 1.

Plasma labels

The most commonly used tracer for plasma is ^{125}I-labelled HSA. This is generally obtained direct from the manufacturer who obtains the HSA from a pool of tested donors; it typically has a shelf life of 1–2 months. The advantages of ^{125}I-HSA over ^{131}I-HSA are, first that its γ-ray energy is easily distinguished from that of ^{51}Cr or ^{99m}Tc, whereas that of ^{131}I is close to that of ^{51}Cr, and secondly that the radiation dose to the patient is usually lower (*see Table 1.3*). Unfortunately, when the HSA is broken down, the uptake by the thyroid of radioiodine from these tracers can result in a fairly high radiation dose to this gland. If it is desired to block this uptake by the thyroid, non-radioactive iodine should be administered to the patient for 3–4 weeks. A suitable dosage is potassium iodide 30 mg/day (Ellis *et al.*, 1977).

Transferrin labelled with ^{113m}In is an alternative plasma label and, because ^{113m}In is obtained from an isotope generator, as is ^{99m}Tc, it can be made available at any hospital despite its short half-life (100 minutes). The short half-life is, in fact, its main advantage as it keeps the radiation dose to the patient very low and allows repeated measurements to be made. The γ-ray energy of ^{113m}In is fairly close to that of ^{51}Cr however, and when ^{113m}In is used ^{99m}Tc should be used as the red cell label rather than ^{51}Cr. The accuracy with which the plasma volume can be measured using ^{113m}In has been questioned, however, by Wootton (1976) who concluded that, although the plasma volume measured using ^{113m}In-transferrin correlated reasonably well with that measured using ^{125}I-HSA in haematologically normal subjects, the accuracy of the measurement is only marginally better than the predicted values obtainable from height and weight considerations (*see* below) and that it is not therefore an adequate substitute for ^{125}I-HSA. ^{59}Fe- and ^{52}Fe-labelled transferrin can also be used to measure the plasma volume, but these radionuclides are not often used for this purpose unless ferrokinetic studies are also being undertaken (*see* Chapter 5).

In summary, ^{125}I-labelled HSA is the tracer most commonly used to label plasma although ^{113m}In may be used if repeat measurements are required. ^{125}I-HSA can be used in conjunction with either ^{51}Cr or ^{99m}Tc as the red cell label, whereas ^{113m}In is best used with ^{99m}Tc. It is important to note, however, that the dilution space of labelled proteins varies with the type of protein labelled and is not necessarily equal to the physiological volume of plasma in circulation. This is discussed in greater detail below. The techniques for labelling transferrin with radioisotopes of indium and iron are given in Appendix 1.

Procedure for blood volume determination

Approximately 10 ml of venous blood is withdrawn from the patient for labelling and, if it is suspected that there might be residual radioactivity in the blood from any previous study using radionuclides, a portion is retained for counting in order that a background correction may be made. If it is not practicable to use the patient's own blood it is essential that the blood used comes from a fully compatible donor. It is also obviously essential that all operations carried out on the blood before it is reinjected are performed under aseptic conditions and that all the solutions and chemicals used are sterile and pyrogen-free. The detailed labelling procedures are given in Appendix 1. When the red cells have been labelled they are generally remixed with the labelled plasma or with saline plus the labelled plasma to ensure that the PCV of the injected suspension is similar to that of the venous blood of the patient. If commercial iodinated HSA is used the quantity added should be 2 kBq/kg body weight. Before reinjection a standard solution must be prepared and the amount injected and its relationship with the standard must be determined. The following two methods are simple and suitable ways of preparing the standard although other methods are equally good (Wright, Tono and Pollycove, 1975; ICSH, 1973).

Method 1. Draw up the labelled blood using the syringe and needle that will be used for the injection and weigh them (w_1). Expel approximately 1 ml into a volumetric flask (eg 250 ml) containing ammonia (11.4 mmol/ℓ) and fill the flask to its mark with further ammonia. Reweigh the syringe and needle (w_2). Inject the labelled blood intravenously, do not flush the syringe, and weigh it again (w_3). The ratio of the standard to the amount injected is then given by $(w_1 - w_2)/(w_2 - w_3)$.

Method 2. Draw up all but about 1 ml of the labelled blood into the syringe that will be used for the injection and draw the remaining blood into a similar syringe (the standard syringe) and make up its volume to that of the first syringe with ammonia solution (11.4 mmol/ℓ). Count both syringes (typically for about 30 seconds each) in a jig positioned about 30 cm from the detector of a scintillation counter and count the background for the same length of time. Provided that the injection syringe is well flushed on injection, the ratio of the standard to the amount injected is given by the ratio of the recorded counts corrected for background. Ideally the syringes should be counted for both radionuclides but provided the suspension is well mixed (as is assumed in *Method 1*) it is sufficient to count using only one energy channel. The contents of the standard syringe are then flushed out into a volumetric flask and made up in the same way as for the first method.

Once the standard and injection have been prepared the patient may be injected with the labelled blood. A stop-watch is started at the mid-point of the injection and venous blood samples of about 20 ml (ideally taken from a vein other than the one used for the injection) are taken approximately 10 and 20 minutes after the injection and the exact time intervals are noted. Further samples are taken at about 40 and 60 minutes if the mixing time is expected to be prolonged (as may be the case in patients with splenomegaly or suffering from shock). The samples are immediately placed into vials containing a solid anticoagulant. Either disodium or dipotassium EDTA (1.25–1.75 mg/ml of blood) or heparin (0.1 mg/ml of blood) are suitable (ICSH, 1973), although because excess EDTA results in an alteration of cell volume and causes a falsely low PCV it is essential, if using EDTA, to keep within the prescribed amounts.

An aliquot of each sample is taken for the measurement of the PCV, and known volumes (typically 4.0 or 5.0 ml) of whole blood are pipetted into counting tubes and each then lysed by the addition of a small quantity of saponin. The remainder of the samples are centrifuged for 10 minutes at about 1000g and identical volumes of plasma pipetted into counting tubes, so that a tube containing whole blood and a tube containing the same volume of plasma are obtained from each sample. Ideally duplicate samples are obtained in each case. A counting tube is filled with the same volume of the standard solution in order to prepare a counting standard, a background sample of pure water or saline is prepared, and finally a sample is prepared of whichever of the two radionuclides being used has the highest γ-ray energy (ie ^{51}Cr if used with ^{125}I, or ^{113m}In if used with ^{99m}Tc). This last sample is necessary for adjusting the counts from the whole blood and standard samples in the lower energy channel to correct for the effect of scattered γ-rays from the higher energy γ-rays of the other radionuclide.

All the samples are then counted. If each sample is counted for about 300 seconds, sufficient counts will usually be obtained for the standard deviation in the count to be less than 2 per cent of its value; normally this is sufficiently precise.

The calculation of the blood volume is performed in the way shown in *Figure 3.1*, which reproduces a typical calculation. The recorded counts are corrected first for background and then for any cross-talk between the two counting channels so that in the example shown a correction is made for the number of counts detected in the ^{125}I channel which are due to the detection of scattered γ-rays from the higher energy γ-ray emission of ^{51}Cr. The red cell volume and plasma volume are then calculated using the following expressions:

Red cell volume

$$= \frac{\text{total counts from entire red cell content of injection}}{\text{counts/ml of red cells}}$$

Plasma volume

$$= \frac{\text{total counts from entire plasma content of injection}}{\text{counts/ml of plasma}}$$

In terms of the quantities calculated from the sample counts these two expressions are given by equations (1) and (2).

Red cell volume (1)

$$= \frac{(^{51}\text{Cr std. count} \times \text{std. volume} \times \text{ratio inj.:std.}) \div (\text{vol. of std. sample})}{(^{51}\text{Cr whole blood sample count}) \div (\text{vol. of sample} \times \text{PCV of sample})}$$

Plasma volume (2)

$$= \frac{(^{125}\text{I std. count} \times \text{std. volume} \times \text{ratio inj.:std.}) \div (\text{vol. of std. sample})}{(^{125}\text{I plasma count}) \div (\text{vol. of plasma sample})}$$

BLOOD VOLUME

SAMPLE	Time of sample	Volume of sample counted	PCV of sample	CHANNEL 1 (125-I)					CHANNEL 2 (51-Cr)			
				Count	Counting time (s)	Counts per second	Corrected for back-ground	Corrected for cross-talk	Count	Counting time (s)	Counts per second	Corrected for back-ground
BACKGROUND	/	/	/	308	600·00	0·513	/	/	890	600·00	1·483	/
^{51}Cr SAMPLE	/	/	/	905	75·21	12·033	11·520	/	20000	75·21	265·922	264·4
STANDARD (dil. to 250 mℓ)	/	3·0	/	20000	102·93	194·307	193·794	8·157 185·6	19410	102·93	188·575	187·1
WHOLE BLOOD (1)	10·0	3·0	0·47	15578	390·93	39·849	39·336	2·166 37·2	20000	390·93	51·160	49·7
WHOLE BLOOD (2)	20·0	3·0	0·47	14845	396·83	37·409	36·896	2·133 34·8	20000	396·83	50·399	48·9
PLASMA (1)	10·0	3·0	/	20000	272·48	73·400	72·9	/	/	/	/	/
PLASMA (2)	20·0	3·0	/	20000	291·12	68·700	68·2	/	/	/	/	/

Ratio injection: standard	4·739		Plasma volume	$(185.6 \times 250 \times 4.739 \times 3.0)/(77.9 \times 3.0)$
Cross-talk correction factor	$\frac{11.520}{264.4} = 0.0436$			$= 2823$ mℓ.
Plasma counts/s at time of injection	77·9		Red cell volume	$(187.1 \times 250 \times 4.739 \times 3.0 \times 0.47)/(49.3 \times 3.0)$
Red blood cells average counts/s	49·3			$= 2113$ mℓ.
			Body PCV	$2113/(2113+2823) = 0.43$
			PCV ratio	$0.43/0.47 = 0.915$

Figure 3.1 The calculation of red cell and plasma volumes from the sample counts

The following three points should be noted:

1. It is not necessary to know the exact volume of each sample counted provided that they are all identical since, if this is the case, the sample volumes in equations (1) and (2) then cancel out.
2. As long as complete mixing of the labelled blood with the unlabelled blood of the circulation occurs within 10 minutes of the injection, both whole blood samples will give approximately the same count for the red cells, and their average value can be used in the calculation. If, however, the mixing time is prolonged, further blood samples should be taken at 40 and 60 minutes after injection, and, if ^{99m}Tc rather than ^{51}Cr is used as the red label, a correction may have to be applied to the measured volume to correct for the elution of ^{99m}Tc from the red cells (*see* Appendix 1).
3. It is nearly always necessary to extrapolate the plasma counts back to the time of injection because of the continuous loss of protein from the plasma pool to the extravascular pool. This should be done assuming an exponential rate of loss and may be simply achieved by plotting the data on semilogarithmic graph paper as shown in *Figure 3.2*.

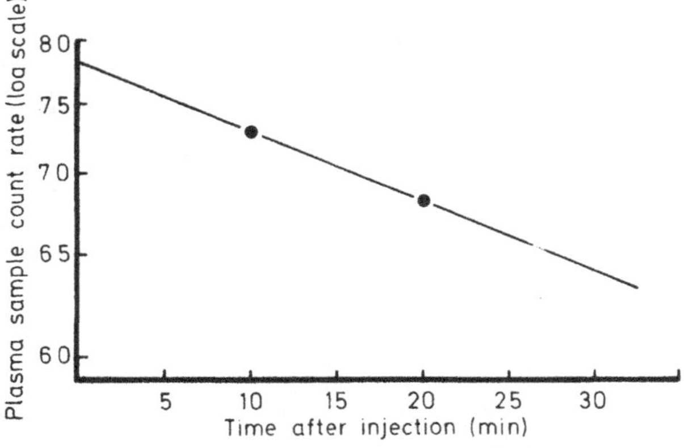

Figure 3.2 The extrapolation of plasma sample counts back to the time of injection. The data shown are used in *Figure 3.1*

Presentation of results and normal values

The red cell and plasma volumes are measured in millilitres, but it is necessary for clinical interpretation to compare the figures with normal reference values and for this purpose they are generally expressed in ml/kg of body weight. Unfortunately, the normal blood volume for any individual is not simple to determine and although many attempts have been made to derive empirical equations based on the height, weight and sex of the subject, none of these are very successful and only measured values which vary by more than about 15–20 per cent from the calculated volume should be considered abnormal. One of the main sources of error in the calculation is the body weight. Since fatty tissue is almost completely avascular the blood volume is more closely related to the lean tissue mass than the total body mass and because of this some authors prefer to express their results as ml/kg lean tissue mass. However, as the accurate determination of lean tissue mass is not practicable on a routine basis the ICSH has recommended that the total body weight should be used, and they have published the normal values shown in *Table 3.1*. No error figure for plasma volume is given as the published results vary much more widely for

Table 3.1 Normal values for red cell and plasma volumes in adults

	Red cell volume (ml/kg)	*Plasma volume* (ml/kg)
Males	25–35	40
Females	20–30	40

plasma than for the red cells, although Dacie and Lewis (1975) give a normal range of 40–50 ml/kg. This variation may be partly due to the fact that the circulating plasma volume is labile and therefore varies with posture and exercise. Because of this the ICSH have recommended (ICSH, 1973) that the patient should be rested in a recumbent position for 15 minutes before the administration of the labelled blood and remain in this position until all the blood samples have been taken. It may also be partly due to the fact that different labelled proteins give different apparent plasma volumes, transferrin gives results about 5 per cent greater than albumin and immunoglobulin results about 5 per cent lower. The albumin volume is usually taken as the standard. Probably the most popular of the empirical equations that are used for the calculation of individual normal blood volume are those of Nadler, Hidalgo and Bloch (1962) and Retzlaff *et al.* (1969), but there are numerous others and as Hurley (1975) has shown none of these can be objectively recommended in preference to any other.

Nevertheless, the practice of comparing the measured volumes with predicted volumes using these empirical equations is becoming increasingly popular and *Table 3.2* and *Table 3.3* can be used to obtain predicted blood volumes as a function of sex, weight and height. These tables are based on the work of Retzlaff *et al.* (1969). *Table 3.2* refers to the red cell volume and *Table 3.3* to the

Table 3.2 The prediction of normal red cell volume. The figures corresponding to the weight and height of the patient added together give the red cell volume in ml.

Weight (kg)	Males	Females	Height (m)	Males	Females
30	519	171	1.45	496	729
35	606	200	1.50	537	811
40	692	228	1.55	578	893
45	779	257	1.60	619	975
50	865	285	1.65	660	1057
55	952	314	1.70	701	1139
60	1038	342	1.75	742	1221
65	1125	371	1.80	783	1303
70	1211	399	1.85	824	1385
75	1298	428	1.90	865	1467
80	1384	456	1.95	906	1549
85	1471	485	2.00	947	1631
90	1557	513			
95	1644	542			
100	1730	570			

Table 3.3 The prediction of normal plasma volume. The figures corresponding to the weight and height of the patient added together give the plasma volume in ml.

Weight (kg)	Males	Females	Height (m)	Males	Females
30	270	252	1.45	1728	1062
35	315	294	1.50	1846	1264
40	360	336	1.55	1965	1467
45	405	378	1.60	2083	1669
50	450	420	1.65	2202	1872
55	495	462	1.70	2320	2074
60	540	504	1.75	2439	2277
65	585	546	1.80	2557	2479
70	630	588	1.85	2676	2682
75	675	630	1.90	2794	2884
80	720	672	1.95	2913	3087
85	765	714	2.00	3031	3289
90	810	756			
95	855	798			
100	900	840			

plasma volume. In both cases it is necessary to add the figure corresponding to the weight of the patient to the figure corresponding to the height of the patient in order to obtain the predicted volume. The functions from which these tables are derived are listed in equations (3) to (6).

Male red cell volume (ml)
$$= 820 \times \text{height (m)} + 17.3 \times \text{weight (kg)} - 693(\pm 252) \qquad (3)$$

Female red cell volume (ml)
$$= 1640 \times \text{height (m)} + 5.7 \times \text{weight (kg)} - 1649(\pm 129) \qquad (4)$$

Male plasma volume (ml)
$$= 2370 \times \text{height (m)} + 9.0 \times \text{weight (kg)} - 1709(\pm 358) \qquad (5)$$

Female plasma volume (ml)
$$= 4050 \times \text{height (m)} + 8.4 \times \text{weight (kg)} - 4811(\pm 196) \qquad (6)$$

The total blood volume is usually taken as the sum of the red cell volume and the plasma volume and therefore ignores the volume of the circulating leukocytes and platelets. Although this does not normally introduce any significant error, in some patients with leukaemia in whom the volume of circulating leukocytes is a significant fraction of the total volume, an underestimate of the total blood volume will result. An estimate of the magnitude of this error may be obtained, however, by comparing the measured PCV with the volume of plasma in the whole blood samples expressed as a fraction of the sample volume.

In order to avoid measuring the red cell and plasma volumes separately the ratio of the body PCV to the venous PCV has been widely used to determine either the plasma volume or the red cell volume from a measurement of the other, and then from this the total blood volume has been deduced. This method relies on the premise that the ratio of PCVs is constant (it is usually taken as being equal to 0.91), but even in normal patients the value of this ratio can vary from one to another by approximately ± 10 per cent (Najean and Cacchione, 1977). Wright, Tono and Pollycove (1975), in one of the largest published series (224 patients), found that the ratio varied from 0.62 to 1.13 with a mean value of 0.89 ± 0.14 (2σ) so that the use of this method can introduce large errors and should not be used. It may however be felt to be worthwhile to calculate the ratio when both PCVs have been determined as part of a blood volume study as long as the large range in normal values is borne in mind. If the measured ratio is higher than normal it is usually taken as being indicative of the presence of a large pool of red cells in the spleen.

In conclusion, blood volume may be measured with considerable precision but it is difficult to be accurate about the expected normal values in any particular patient and its clinical usefulness is therefore less than might be expected. However, when serial measurements are required in any one patient, changes in blood volume of the order of a few per cent are readily detectable. The clinical use of the measurement in many disease states has been comprehensively reviewed by Najean and Cacchione (1977). The measurement is particularly useful in polycythaemia where the PCV alone has been shown to be a poor index of the severity of the disease (Szur, Lewis and Goolden, 1959).

References

DACIE, J. V. and LEWIS, S. M. (1975) *Practical Haematology*, 5th Edn. p.442. Edinburgh and London: Churchill Livingstone

ELLIS, R. E., NORDIN, B. E. C., TOTHILL, P. and VEALL, N. (1977) The Use of Thyroid Blocking Agents – A report of a working party of the MRC Isotope Advisory Panel. *British Journal of Radiology*, **50**, 203–204

FERRANT, A., LEWIS, S. M. and SZUR, L. (1974). The elution of ^{99m}Tc from red cells and its effect on red cell volume measurement. *Journal of Clinical Pathology*, **27**, 983–985

GLASS, H. I., BRANT, A., CLARK, J. C., de GARRETA, A. C. and DAY, L. G. (1968) Measurement of blood volume using red cells labelled with radioactive carbon monoxide. *Journal of Nuclear Medicine*, **9**, 571–575

HURLEY, P. J. (1975) Red cell and plasma volumes in normal adults. *Journal of Nuclear Medicine*, **16**, 46–52

ICSH (1973) Standard techniques for the measurement of red cell and plasma volume. *British Journal of Haematology*, **25**, 801–814

JONES, J. and MOLLISON, P. L. (1978) A simple and efficient method of labelling red cells with ^{99m}Tc for determination of red cell volume. *British Journal of Haematology*, **38**, 141–148

KORUBIN, V., MAISEY, M. N. and MACINTYRE, P. A. (1972) Evaluation of technetium labelled red cells for determination of red cell volume in man. *Journal of Nuclear Medicine*, **13**, 760–763

MOLLISON, P. L., ROBINSON, M. and HUNTER, D. (1958) Improved method of labelling red cells with radioactive phosphorus. *Lancet*, **i**, 766–774

NADLER, S. B., HIDALGO, J. V. and BLOCH, T. (1962) Prediction of blood volume in normal human adults. *Surgery*, **51**, 224–232

NAJEAN, Y. and CACCHIONE, R. (1977) Blood volume in health and disease. *Clinics in Haematology*, **6**, 543–566

RETZLAFF, J. A., TAUZE, W. N., KIELY, J. M. and STROEBAL, C. F. (1969) Erythrocyte volume, plasma volume and lean body mass in adult men and women. *Blood*, **33**, 649–667

SZUR, L., LEWIS, S. M. and GOOLDEN, A. W. G. (1959) Polycythaemia vera and its treatment with radioactive phosphorus: a review of 90 cases. *Quarterly Journal of Medicine*, **28**, 397–412
WOOTTON, R. (1976) The limitations of ^{113m}In for plasma volume measurement. *British Journal of Radiology*, **49**, 427–429
WRIGHT, R. R., TONO, M. and POLLYCOVE, M. (1975) Blood Volume. *Seminars in Nuclear Medicine*, **5**, 63–78

4

Cell survival studies

Studies of blood cell survival in the circulation have been concentrated on the study of red cells although a few centres have routinely measured platelet and white cell survival in addition. These latter studies have not yet become universally used in the same way as the red cell studies. This is reflected by the relative amounts of space devoted to each of these studies in this chapter, although much of the information included under the red cell studies is also directly applicable to the measurement of platelet and white cell survival. If the development of the new ^{111}In labelling techniques for platelets and white cells lives up to its promise, these latter studies could well become much more widely used in the future.

Measurement of red cell survival

Red-cell survival studies were first successfully performed by Ashby (1919) who administered donor cells of a different but compatible blood group to the patient under investigation. Blood samples were then taken and the mixture of cells separated *in vitro* by an antiserum that agglutinated the recipient's red cells but not the donor cells, and the donor cells were then counted. This method, although used for several decades, has gone out of general use and has been replaced by radionuclide tracer techniques.

Two broadly separate techniques have been used with radionuclides; first the so-called 'cohort' labelling techniques in which red cells produced in the marrow over a short period of time only are labelled, and secondly the random labelling techniques in which samples of circulating cells of all ages are labelled.

The traditional cohort labelling techniques involve the administration of a tracer which is incorporated into the red cells during their formation. The labelled cells are then released into the blood stream as a cohort of cells of similar age and the amount of tracer circulating in the blood stream is monitored. The theoretical curve from an ideal study shown in *Figure 4.1* demonstrates the rise in blood radioactivity as the cohort of labelled cells is released into the circulation and the subsequent fall when the cells reach the end of their lifespan. Three nuclides have been used for these studies, ^{14}C and ^{15}N labelled glycine and ^{59}Fe, although each of these

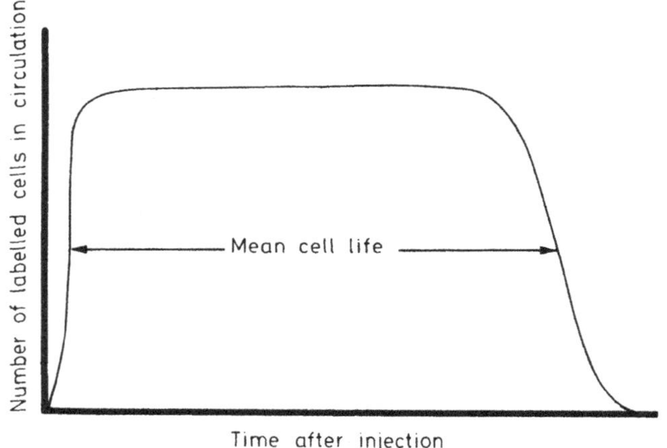

Figure 4.1 An ideal curve obtained from a cohort study of red cell survival in a normal subject

tracers has disadvantages. Glycine does not reach its maximum level in the circulation for about 25 days so that the leading edge of the curve is not well defined and iron is so effectively reused by the body that the fall in the curve at the end of the red cell lifespan is normally masked. In addition to these problems it is necessary to make measurements over at least 120 days in normal studies and this makes the test very time consuming. Because of these practical difficulties the techniques in normal clinical use are the random labelling techniques, although methods of obtaining the mean cell life from ^{59}Fe measurements have recently been developed which do not require the complete curve to be obtained. These are discussed with other ferrokinetic studies in Chapter 5.

The two compounds which have been universally used in random labelling of red cells for survival measurement are ^{51}Cr sodium chromate and di-isopropylfluorophosphonate (DFP) labelled with either ^{32}P or ^{3}H. Both these compounds have advantages and disadvantages with respect to each other and these are discussed below. If the survival of the cells is to be determined accurately using a random label, the label must satisfy certain requirements:

1. The label must be attached to the cells in a truly random manner.
2. It must not be toxic to the cells.
3. It must remain firmly bound to the cells for their entire lifetime.
4. It must not relabel new cells when cells to which it was originally bound are broken down.

In addition to these there are also the more technical requirements that the labelling procedure, sample preparation and counting procedures should be as straightforward as possible. Unfortunately, neither of the two labels available meet all these requirements although of the two ^{51}Cr is the more widely used.

The primary advantage of ^{51}Cr sodium chromate over the DFP labels is that ^{51}Cr emits γ-rays while ^{32}P and ^{3}H only emit β-particles. Sample preparation and counting are much simpler therefore, and it is also possible with ^{51}Cr to perform external counting over the liver and spleen in conjunction with the cell survival measurements in order to obtain information about the sites of red cell destruction. (This aspect is discussed further in Chapters 6 and 7.) The primary disadvantage of ^{51}Cr is that it does not remain firmly bound to the labelled cells but is slowly eluted from them, the rate of loss being of the order of one per cent per day, and this makes a precise determination of the mean cell life more difficult and less accurate than it is with DFP where the elution problem is greatly reduced.

Obtaining the survival curve using ^{51}Cr

The technique for labelling the cells with ^{51}Cr is the same as that used for the measurement of blood volume but it is not necessary in this case to prepare any standards unless the red cell volume is to be determined simultaneously. The cells are normally labelled with 25–50 kBq of ^{51}Cr/kg body weight, the higher quantity being used if external counting measurements are also to be made.

The initial blood sample should be taken 10 minutes after injection of the labelled cells, or 60 minutes if mixing is expected to be slow. Ideally it should be taken from a vein other than the one used for the injection. A further sample should be taken at 24 hours and then at least 3 samples/week for the rest of the study (generally 2–3 weeks). The samples should be taken into an anticoagulant in the same way as for the blood volume studies described in Chapter 3 (*see* page 35). In order to prepare the samples for counting the cells should be lysed by the addition of a small quantity of solid saponin to the sample and be thoroughly mixed. Identical volumes are then pipetted from each sample into counting tubes and the radioactivity of each sample measured.

Traditionally the data are then plotted and an index known as the T_{50}Cr or $T_{1/2}$Cr obtained, this being the time taken for half of the injected ^{51}Cr to be lost from the circulation. The normal range of the T_{50}Cr is 25–33 days (Dacie and Lewis, 1975). Unfortunately, because of the steady elution of the label from the red cells and because the disappearance of the red cells from the circulation may have more than one component this index does not bear a simple relationship to the mean red-cell life which is the parameter usually required in clinical practice. Because of this the ICSH has recommended (ICSH, 1971) that the use of this index be abandoned and that, as far as possible, the data be corrected for elution in order than a mean cell life can be calculated. To this end, elution correction tables have been prepared for the two most commonly used ^{51}Cr labelling techniques and these are given in Appendix 1 with the labelling methods. In order to correct for elution the uncorrected counts are multiplied by the elution correction factors. The sample counts are then expressed as a percentage of the initial sample counts using the expression:

$$\text{percentage survival on day } N = \frac{100 \times \text{counts/ml on day } N \times \text{elution correction factor}}{\text{counts/ml on day } 0}$$

These results are then plotted and the mean red-cell life obtained from the curve (*see* below). The calculation of the percentage survival from the raw counting data is shown in *Table 4.1*.

There is normally no necessity to correct the measurements for variations in the PCV of the samples as long as the total blood volume remains constant throughout the study. This is sometimes not the case and then this factor may limit the accuracy of the measurement. If the blood volume is known to be changing, serial blood volume determinations may be carried out during the study using ^{99m}Tc and ^{113m}In as described in Chapter 3. The elution-corrected percentage survival on any day should then be multiplied by the ratio of the blood volume on that day to the blood volume on the first day of the study in order to obtain the correct survival. If it is not practicable to do this it is important to bear in

Table 4.1 Calculation of the percentage survival from ^{51}Cr sample counting data

Day of study	Counts recorded	Counting time (s)	Counts/s	Corrected for background (counts/s)	Elution factor	Corrected for elution	Percentage survival
(background)	902	600.00	1.50	–	–	–	–
0	20 000	149.26	133.99	132.49	1.00	132.49	100.0
1	20 000	153.63	130.18	128.68	1.01	129.97	98.1
3	20 000	161.55	123.80	122.30	1.04	127.19	96.0
5	20 000	167.53	119.38	117.88	1.07	126.13	95.2
7	20 000	177.89	112.43	110.93	1.10	122.02	92.1
10	20 000	187.89	106.45	104.95	1.14	119.64	90.3
12	20 000	197.28	101.38	99.88	1.17	116.86	88.2
14	20 000	205.94	97.12	95.62	1.20	114.74	86.6
17	20 000	221.77	90.18	88.68	1.24	109.96	83.0
19	20 000	229.21	87.26	85.76	1.27	108.92	82.2
21	20 000	243.28	82.21	80.71	1.31	105.73	79.8

mind this source of error. In practice, if for instance a patient needs to be transfused during a study, it can normally be assumed that the total blood volume will have reverted to its pre-transfusion level after 48 hours.

Obtaining the survival curve using DFP

Labelling with DFP can be carried out either *in vitro* or *in vivo* but is normally performed *in vivo* as a large volume of cells (100–200 ml) is required for *in vitro* labelling. For *in vivo* labelling the DFP solution is slowly injected intravenously over a period of 10–15 minutes. The quantity of DFP injected should not exceed 0.02 mg/kg body weight since it is toxic in greater quantities and the radioactivity injected should not be greater than 25 kBq/kg body weight for ^{32}P-DFP or 250 kBq/kg body weight for ^{3}H-DFP (of this radioactivity 30–40 per cent only is taken up by the red cells). The labelled DFP is commercially available.

Blood samples are taken in the same way as for the ^{51}Cr method except that the first sample is always taken at 60 minutes in order to allow time for complete labelling to have occurred. Each sample must be carefully washed at least three times with isotonic saline before it is prepared for counting because the DFP also labels circulating leukocytes and platelets and these must be removed from the sample before counting. Since both ^{32}P and ^{3}H are pure β-particle emitters, sample preparation for counting is more complex that it is for ^{51}Cr. ^{32}P samples are commonly counted with a very thin window Geiger counter or in a gas flow Geiger counter, and ^{3}H samples (because of their lower β-particle energy) in a liquid scintillation counter. For Geiger counter measurements precise sample preparation details depend on the detailed design of the instrument but usually a known volume of haemolysed red cells is dried on a tray or planchet which is then put into the counter. For liquid scintillation counting it is normal to burn a known weight of dried haemolysed cells in a combustion apparatus and then count the tritiated water that is produced, but again the exact details depend on the combustion apparatus and on the liquid scintillants available. It is not normally possible to count a solution of haemolysed cells directly by liquid scintillation counting as the colour quenching (*see* Chapter 2) is too severe. Finally the results are plotted in the same way as for ^{51}Cr except that no elution correction factors are required. Some elution of the label during the first 24 hours of the study has been reported (Bove and Ebaugh, 1958; Eernisse and Van Rood, 1961) but the reason for this has not been satisfactorily explained. Mollison (1979) has

suggested that it may be due to technical factors and Bentley *et al.*
(1974) failed to demonstrate any significant elution of the label at
all. However, if the initial point on the survival curve does lie
significantly higher than would be expected from back extrapolation of the subsequent data, the point can be ignored and the
extrapolated value used as the initial point.

As with the ^{51}Cr measurements, if the blood volume is known to
be changing during the study, a correction should be made to the
measured survival. In this case, since the samples counted are red
cells alone rather than whole blood, it is only necessary to make
serial determinations of the red cell volume. The measured
survival on any day should be multiplied by the ratio of the red cell
volume on that day to the red cell volume on the first day of the
study.

Interpretation of the survival curve and calculation of the mean cell life

The theory of the shape of the red cell survival curve was worked
out by Dornhorst (1951). The theory assumes that the cells can be
destroyed by one or both of two mechanisms; first by a random
destruction mechanism where each cell has the same probability of
being destroyed in unit time irrespective of its age, and secondly
by a senescence mechanism in which cells are destroyed when they
reach a certain age. The general form of the survival curve is given
in equation (1), where $P(t)$ is the percentage of the labelled cells
surviving at time t after injection, k is the rate of random
destruction and T is the age at which death from senescence
occurs. The values of k and T vary from patient to patient
depending on the relative amounts of destruction occurring due to
each mechanism and therefore the shape of the curve also varies.
Since the original derivation of this equation is somewhat obscure
an alternative derivation of it is given in Appendix 2.

$$P(t) = 100\,(e^{-kt} - e^{-kT})/(1 - e^{-kT}) \tag{1}$$

In practice it is often found that one or other mechanism of
destruction dominates and the equation can then be simplified. If
all destruction is random equation (1) is reduced to the simple
exponential expression given in equation (2).

$$P(t) = 100e^{-kt} \tag{2}$$

If there is no random destruction and all cell death is due to
senescence equation (1) is reduced to the linear function of time

given in equation (3). (Both these simplified expressions are also derived in Appendix 2.)

$$P(t) = 100\,(1 - t/T)\tag{3}$$

Thus, if the survival data is plotted on both linear and semilogarithmic paper a straight line will be obtained on the linear plot only if death is purely from senescence and on the semilogarithmic paper only if death is purely due to random destruction. It is therefore possible by simple visual inspection of the two curves to obtain some information about the mechanism or mechanisms of destruction. In the general case, in which both destruction mechanisms are acting, a straight line will not be obtained on either plot.

The mean cell life (MCL) is given by the reciprocal of the initial rate of destruction. For the general case given in equation (1) this is given by expression:

$$\mathrm{MCL} = (1 - e^{-kT})/k\tag{4}$$

For the case of purely random destruction it is given by the expression:

$$\mathrm{MCL} = 1/k\tag{5}$$

and for the case of death purely from senescence by the simple expression:

$$\mathrm{MCL} = T\tag{6}$$

All these expression are also derived in Appendix 2.

Values for the MCL can be simply obtained from the survival curves by using graphical procedures. If the data form a straight line on the linear plot the MCL is given by the time at which there would be no cells surviving, ie the time at which a line drawn through the points cuts the time axis. If, however, the data form a straight line on the semilogarithmic plot the MCL is given by the time at which 37 per cent of the cells are still surviving. If a straight line cannot be achieved on either plot, thus showing that both destruction mechanisms are present, an estimate of the MCL can still be obtained from the linear plot by drawing a tangent to the initial slope of the curve and taking the value where this cuts the time axis. Alternatively, and more accurately, the data may be

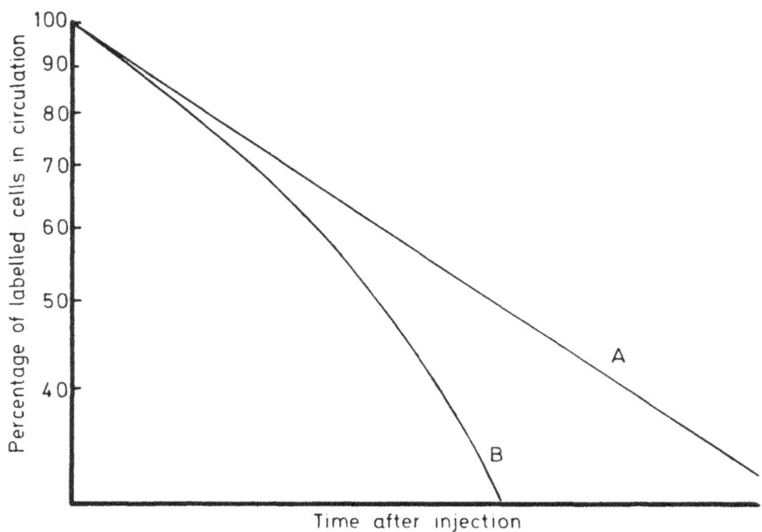

Figure 4.2 Typical curves of red cell survival plotted on semilogarithmic graph paper: (A) cell destruction by a random mechanism, and (B) cell destruction by a senescence mechanism

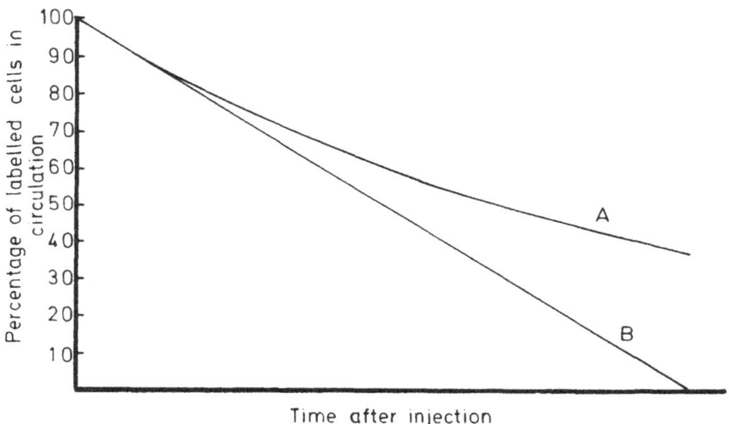

Figure 4.3 Typical curves of red cell survival plotted on linear graph paper: (A) cell destruction by a random mechanism, and (B) cell destruction by a senescence mechanism

fitted by computer to equation (1) using a least squares method to get the best fit to the data, and values for k and T may then be obtained. These values are then substituted into equation (4) to obtain the MCL. Typical survival curve shapes are shown in *Figures 4.2* and *4.3* where curve A is pure random destruction and curve B is due to senescence only. The normal MCL is 110 days and this has a standard deviation of 20 days (Bentley *et al.*, 1974).

In some anaemias (typically sickle cell disease and paroxysmal nocturnal haemoglobinuria), and sometimes in patients in whom a mixture of transfused normal cells and the patient's own cells has been labelled, the survival curve consists of two components due to two populations of cells that have different MCLs. In this case it is necessary to separate the survival curve into its two components before analysis. For this to be done accurately it is essential to obtain more samples than is usually the case and daily blood samples should be taken throughout the first week after injection at least. The separation of the two components is achieved by extrapolating the tail of the curve back to the time of injection in order to obtain the percentage of the longer lived component and then substracting this from the original data to obtain the percentage of the shorter lived component. Each component is then separately analysed as shown in *Figure 4.4* where it can be

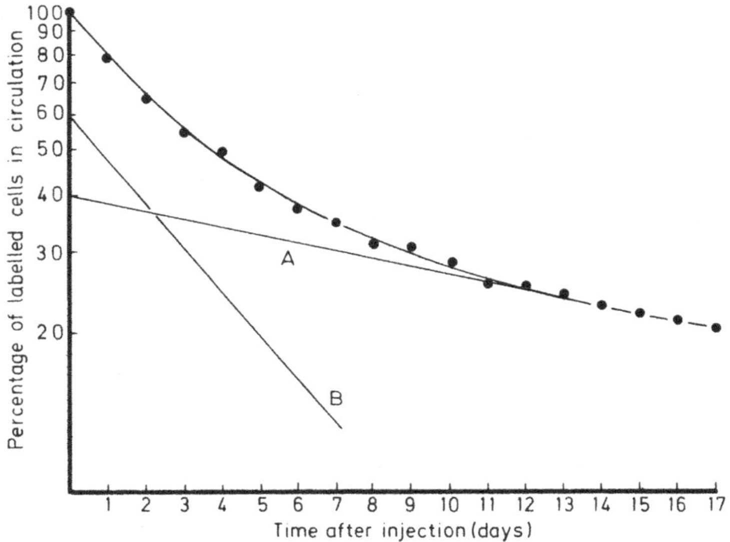

Figure 4.4 A red cell survival curve plotted on semilogarithmic graph paper showing a double population of cells. Approximately 40 per cent of the cells belong to the longer-lived population (A), and the remaining 60 per cent to the shorter-lived population (B)

seen that the longer lived component accounts for 40 per cent of the total population and the shorter lived component for the remaining 60 per cent. If the MCL for each component is obtained the MCL of the total population can be obtained from the expression given in equation (7), where L_1 and S_1 are the mean cell life and percentage abundance of one population and L_2 and S_2 are the mean cell life and percentage abundance of the other population.

$$MCL = 100/[(S_1/L_1) + (S_2/L_2)] \tag{7}$$

Thus, in the example illustrated in *Figure 4.4*, where the short-lived population has a MCL of 4.5 days and the longer-lived population has a MCL of 26 days, the overall MCL is just under 9 days.

Another complication occurs if it is known that there is a steady loss of blood from the patient during the study, perhaps due to bleeding from the gastrointestinal tract or even to the taking of an excessive number of blood samples. Such a loss can have the effect of artificially shortening the measured mean cell life. A corrected value can be obtained using equation (8), where L is the apparent mean cell life, RCV is the red cell volume and V is the volume of red cells lost/day.

$$MCL = L \times RCV/[RCV - (L \times V)] \tag{8}$$

This correction is generally only important if more than 20 ml of blood is being lost from the patient each day.

Choice of method

As previously discussed, neither sodium chromate nor DFP are ideal compounds for labelling red cells. ^{51}Cr is not ideal because of its elution and also because there is some evidence that younger cells are labelled preferentially (Bentley, 1977). DFP is not ideal because both its labels are β-emitters, sample preparation and counting are relatively involved procedures, and it is also not possible to perform external counting measurements.

The error which is introduced by the elution of ^{51}Cr is often underestimated as there are substantial variations in elution rate from patient to patient and the standard elution correction tables use average values from normal subjects. In one series of haematologically normal patients the range was from 0.7–1.6 per cent/day (Bentley *et al.*, 1974) and there is some evidence that the variability may be even greater in patients with haematological disorders (Cline and Berlin, 1963; Hutcheon *et al.*, 1977). However, in clinical practice, red cell survival studies are usually carried out to assess the significance of a known haemolytic process, and clinically acceptable estimates of the mean cell life are generally achieved with ^{51}Cr. The DFP methods, therefore, are normally only used when a higher degree of accuracy is required, such as when trying to measure the small amounts of red cell destruction which may be masked by the errors inherent in the chromium elution corrections.

The methods available for measuring the mean red cell life using ^{59}Fe which are discussed in the next chapter have the advantage that there is no elution of the label, but they give no information on the shape of the survival curve and cannot be used, as ^{51}Cr can be used, to obtain information by external counting about the relative amounts of destruction occurring in each organ. If ferrokinetic studies are being performed on the patient, however, these methods do provide a measurement of the mean red cell life with no additional radiation dose to the patient.

Measurement of platelet survival

The measurement of platelet survival has proved of value in studies of the mechanisms of thrombocytopenia and in investigations of the effects of different therapies on platelet survival. It has not, however, become a routine investigation in more than a few centres and will therefore be only briefly discussed. Its lack of use is due primarily to the technical difficulties in labelling the platelets, since these cells are very susceptible to damage on handling and also have an inherent tendency to adhere both to foreign surfaces and to one another.

Choice of tracer

Various radionuclides have been used to label platelets but, as with red cells, only ^{32}P-DFP and ^{51}Cr sodium chromate have been at all widely used (Szur, 1971; ICSH, 1977). The advantage of DFP is that it is an *in vivo* label but the survival curves obtained with it never reach zero and generally plateau at approximately 15 per cent of the initial level. The cause of this has not yet been fully elucidated but is partly due to reutilization of the label. In addition, since red and white cells take up the label to a greater extent than the platelets, it is necessary to isolate the platelets from each sample before preparing the samples for counting. Because of these problems with DFP, ^{51}Cr is, at present, the standard radionuclide with which to label platelets. The method recommended by ICSH can be used to label autologous platelets even in cases of severe thrombocytopenia (ICSH, 1977). The recommended labelling technique is described in Appendix 1, but although this method minimizes damage to the platelets during labelling, it is not known whether any significant elution of the label occurs after injection of the labelled platelets into the circulation nor whether younger cells are preferentially labelled (Tessier, Steiner and Baldini, 1974).

Recently, ^{111}In has also been successfully used as a label for platelets and results comparable with those from ^{51}Cr have been obtained (Goodwin *et al.*, 1978; Heaton *et al.*, 1979). The labelling procedure with ^{111}In is more efficient than with ^{51}Cr so that much smaller volumes of blood are required and the time required for labelling is less. In addition, the physical characteristics of ^{111}In (the shorter half-life and better γ-ray emission—*see Table 1.2*) are better than those of ^{51}Cr and sufficient radioactivity can be used to enable the distribution of the labelled platelets in the body to be visualized with radionuclide imaging equipment. Because of this, the labelling method used by Heaton *et al.* (1979) is also described in Appendix 1 although as it is still in the evaluation phase it cannot yet be regarded as a standard method. It may be expected when the labelling techniques with this radionuclide have been standardized that it will become the label of choice.

Method

When the patient's platelets have been labelled with ^{51}Cr or ^{111}In, the following procedure should be followed. After injection of the labelled platelets blood samples are taken at 1 hour and 4 hours, and then daily at the same time, if possible, each day. The normal mean cell life is 9–10 days (Harker, 1977) but if the survival is expected to be markedly reduced additional samples should be taken on the first and second days of the study. For each sample, 10 ml of blood are normally required and these should be taken into a syringe containing 0.2 ml of 100 g/ℓ EDTA. The platelet count is determined from an aliquot of the blood, and a measured volume of the sample is then pipetted into a plastic test tube and diluted with an equal volume of saline. This is then centrifuged at 300 g for 10 minutes after which the platelet rich plasma (PRP) is removed. The centrifugation is then repeated and the two harvests of PRP are pooled. The platelet count of the pooled PRP is determined, the volume accurately measured and the solution placed in a counting vial. The vial containing PRP is centrifuged at 2000g for 30 minutes and the supernatant removed without disturbing the platelet button. The sample is then ready for counting. The corrected platelet radioactivity (P) at the time of sampling is given by the expression in equation (9), where Q is the measured radioactivity in the platelet button, N_1 is the platelet count in the original whole blood sample, N_2 is the platelet count in the pooled PRP, V_1 is the volume of whole blood from which the PRP was prepared and V_2 is the volume of the pooled PRP.

$$P = (Q \times N_1 \times V_1)/(N_2 \times V_2) \tag{9}$$

This quantity is then plotted against time in order to obtain the survival curve. The survival curve is usually a mixture of linear and exponential functions and the same techniques can be used to obtain the mean lifespan as are used for red cell survival studies (ICSH, 1977). However, the initial data are often subject to error due to cell damage during the labelling process and, unless the data form a good curve, these methods are not particularly reliable.

An alternative method for obtaining the mean lifespan has been developed by Murphy and Francis (1971) in which the data are fitted by computer to a gamma function. This method has not achieved widespread use although good experience of it has been reported (Harker, 1977) and it has been recommended by the ICSH (1977) for use when the data are not sufficiently good for the simpler methods to be used or when maximum precision is required. The method is based on a 'multiple hit' model in which it is assumed that a platelet can be 'hit' a certain number of times (n) before it is destroyed. If the reciprocal of the mean time between hits is designated A the mean survival time is given by n/A. A computer is used to find the values of n and A that minimize the difference between the observed and calculated radioactivity in the samples. The calculated radioactivity in the samples at time t after injection [$H(t)$] is obtained by computer from the function in equation (10) in which C is the initial level of radioactivity at the time of injection (ICSH, 1977).

$$H(t) = \frac{C}{n} \sum_{i=0}^{n-1} \left[\frac{(n-i)}{i!} . e^{-At} . (At)^i \right] \tag{10}$$

Measurement of white cell survival

Survival studies with white cells have been almost entirely limited to measurements with neutrophils and lymphocytes, but both these tests are still very much in the development stage and it is not possible, at present, to assess their use in clinical haematology. Neutrophil kinetics have been recently reviewed by Vincent (1977), but since then new labelling techniques have been developed using ^{111}In (Thakur, Coleman and Welch 1977; Lavender *et al.*, 1977; Segal *et al.*, 1978) and, although the use of these has not yet become very widespread, once a standard labelling technique has been agreed (as with platelets) these studies may be expected to be much more widely performed.

References

ASHBY, W. (1919) The determination of the length of life of transfused blood corpuscles in man. *Journal of Experimental Medicine,* **29**, 267

BENTLEY, S. A., GLASS, H. I., LEWIS, S. M. and SZUR, L. (1974) Elution correction in ^{51}Chromium red cell survival studies. *British Journal of Haematology,* **26**, 179–184

BENTLEY, S. A. (1977) Red cell survival studies reinterpreted. *Clinics in Haematology,* **6**, 601–623

BOVE, J. R. and EBAUGH, F. G. (1958) The use of di-isopropyl-fluorophosphonate^{32} for the determination of *in vivo* red cell survival and plasma cholinesterase turnover rates. *Journal of Laboratory and Clinical Medicine,* **51**, 916

CLINE, M. J. and BERLIN, N. I. (1963) The red cell chromate elution rate in patients with some haematological diseases. *Blood,* **21**, 63

DACIE, J. V. and LEWIS, S. M. (1975) *Practical Haematology,* 5th Ed. p.461. Edinburgh and London: Churchill Livingstone

DORNHORST, A. C. (1951) The interpretation of red cell survival curves. *Blood,* **6**, 1284

EERNISSE, J. G. and VAN ROOD, J. J. (1961) Erythrocyte survival time deterinations with the aid of DF^{32}P. *British Journal of Haematology,* **7**, 382

GOODWIN, D. A., BUSHBERG, J. T., DOHERTY, P. W., LIPTON, M. J., CONLEY, F. K., DIAMANTI, C. I. and MEARES, C. F. (1978) Indium-111 labelled platelets for localisation of vascular thrombi in humans. *Journal of Nuclear Medicine,* **19**, 626–634

HARKER, L. A. (1977) The kinetics of platelet production and destruction in man. *Clinics in Haematology,* **6**, 671–693

HEATON, W. A., DAVIS, H. D., WELCH, M. J., MATHIAS, C. J., JOIST, J. H., SHERMAN, L. A. and SIEGAL, B. A. (1979) Indium-111. A new radionuclide label for studying human platelet kinetics. *British Journal of Haematology,* **42**, 613–622

HUTCHEON, A. W., HORTON, P. W., ORR, J. S. and DAGG, J. H. (1977) The assessment of red cell survival in normal subjects and in patients with haemolytic disorders and ineffective erythropoiesis using the radioiron ocupancy method. *British Journal of Haematology,* **37**, 195–205

ICSH (1971) Recommended methods for radioisotope red cell survival studies. *British Journal of Haematology*, **21**, 241–250

ICSH (1977) Recommended methods for radioisotope platelet survival studies. *Blood*, **50**, 1137–1144

LAVENDER, J. P., GOLDMAN, J. M., ARNOT, R. N. and THAKUR, M. L. (1977) Kinetics of Indium-111 labelled lymphocytes in normal subjects and patients with Hodgkins disease. *British Medical Journal*, **2**, 797–799

MOLLISON, P. L. (1979) *Blood transfusion in clinical medicine*, 6th Edn. p. 29. Oxford: Blackwell Scientific Publications

MURPHY, E. A. and FRANCIS, M. E. (1971) The estimation of platelet survival. The multiple hit model. *Thrombosis et Diathesis Haemorrhagica*, **25**, 53–80

SEGAL, A. W., DETEIX, P., GARCIA, R., TOOTH, P., ZANELLI, G. D. and ALLISON, A. C. (1978) Indium-111 labelling of leukocytes: A detrimental effect on neutrophil and lymphocyte function and an improved method of cell labelling. *Journal of Nuclear Medicine*, **19**, 1238–1244

SZUR, L. (1971) Blood cell survival studies. In *Radioisotopes in Medical Diagnosis*. pp. 342–382. Eds. Belcher, E. H. and Vetter, H. London: Butterworths

TESSIER, C., STEINER, M. and BALDINI, M. G. (1974) Measurement of platelet kinetics using ^{51}Cr. In *Platelets, Production, Function, Transfusion and Storage*. p. 327. Eds. Baldini, M. G. and Ebbe, S. New York: Grune and Stratton.

THAKUR, M. L., COLEMAN, R. F. and WELCH, M. J. (1977) ^{111}In labelled leukocytes for the localisation of abscesses. Preparation, analysis, tissue distribution and comparison with ^{67}Ga citrate in dogs. *Journal of Laboratory and Clinical Medicine*, **89**, 217–228

VINCENT, P. C. (1977) Granulocyte kinetics in health and disease. *Clinics in Haematology*, **6**, 695–717

5

Ferrokinetics

Studies with radioisotopes of iron were one of the first applications of radionuclide tracer techniques in haematology (Hahn *et al.*, 1939) and they remain important today. Traditionally three parameters have been measured: the half-clearance time from the plasma of intravenously injected iron; the plasma iron turnover (PIT); and the percentage of the injected iron utilized by the red cells (the iron utilization). From these parameters it is possible to obtain an approximate measure of the amount of erythropoiesis occurring in the patient. Unfortunately, there is a large spread in the normal values obtained from these measurements and this reduces their clinical usefulness. The value of the simple measurements in routine clinical practice has been questioned, therefore, and several attempts have been made to interpret the measurements in terms of compartmental models of iron physiology in order that more specific results may be obtained. In the past these have not proved particularly useful clinically and have been used primarily in research investigations. More recently, however, some success has been achieved and ferrokinetic measurements have been successfully used to measure red cell survival. Some of these measurements are discussed in the latter half of this chapter. In addition to the ferrokinetic measurements, the movement of the iron is often followed after injection by external detectors; this is discussed in Chapter 6. The measurement of iron absorption is discussed in Chapter 8.

Three radioisotopes of iron have been used for these studies; ^{59}Fe, ^{55}Fe and ^{52}Fe. The most commonly used and widely available of these radionuclides is ^{59}Fe, which has a radioactive half-life of 45 days and emits high-energy gamma rays. In the past ^{55}Fe was used a great deal but, because it has a 3 year half-life and emits only very low energy photons, it is not now in widespread use. The third isotope, ^{52}Fe is a cyclotron-produced radionuclide with a half-life of 8.2 hours which decays by positron emission and emits 511 MeV annihilation radiation photons. The relatively short half-life of ^{52}Fe has two effects, first it means that this radionuclide can only be used for the initial phases of a ferrokinetic study, and secondly it means that larger quantities can be given to a patient than is possible with the other two isotopes. Because of this it is possible to administer sufficient ^{52}Fe to visualize the distribution of functioning reticuloendothelial tissue a few hours after injection; this is discussed further in Chapter 7. In general ^{52}Fe is used only if such imaging is to be done.

Plasma clearance and plasma iron turnover

Plasma clearance

In order to obtain a satisfactory clearance curve it is necessary to ensure that the administered iron is bound to transferrin in the plasma. Normally there will be sufficient unsaturated transferrin in the plasma for this not to be a problem, but if there is any doubt about it the iron binding capacity of the patient's plasma should be determined (ICSH, 1978), and, if less than $0.5\,\text{mg}/\ell$, normal donor plasma should be used instead (Dacie and Lewis, 1975). The labelling procedure is given in Appendix 1. A standard must be prepared from the labelled transferrin using a method similar to one of those described for the preparation of the standard in the measurement of plasma volume (*see* Chapter 3) except that the standard should be diluted with 0.1M HCl.

After injection of the labelled plasma, blood samples are taken at approximately 5, 10, 20, 40, 60, 90, 120 and 180 minutes, and the exact times noted. The blood should be taken from a vein other than the one used for the injection, and collected into solid heparin or EDTA as described in Chapter 3 for blood volume measurements. The samples are centrifuged at $2000\,g$ for 5 minutes and the plasma removed. The radioactivity per unit volume of plasma is then determined for each sample and the results plotted on semilogarithmic graph paper. Normally a straight line can be accurately fitted to the data in the initial part of the curve and the half-clearance time obtained from this. The most accurate way of obtaining the half-clearance, however, is to fit the data to the simple exponential function given in (1) using a least squares method—most programmable calculators can do simple fits of this kind.

$$A_t = A_0\,e^{-0.693t/T_{1/2}} \tag{1}$$

In this expression A_t is the measured counts/second/ml of plasma from the sample taken at time t after injection, $T_{1/2}$ is the half-clearance time and A_0 is the value of A_t at the time of injection assuming complete instantaneous mixing. The values of A_0 and $T_{1/2}$ are obtained from the calculator. A typical clearance curve is shown in *Figure 5.1*. A portion of at least one of the plasma samples must be retained for the chemical measurement of the total plasma iron (ICSH, 1978).

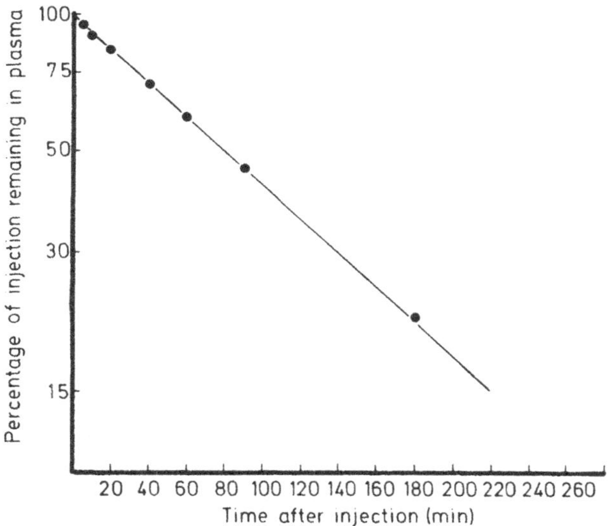

Figure 5.1 The clearance of ^{59}Fe from plasma in a normal subject following intravenous injection. The data plotted show a $T_{1/2}$ of 80 min and this value is used in *Table 5.1*

In most cases the clearance rate depends primarily on the amount of erythopoiesis occurring but the clearance rate also depends on the level of plasma iron and is prolonged if the plasma iron is high. A rapid clearance is usually found in iron deficiency and haemolytic anaemias and a slow clearance in aplastic anaemias. The normal half-clearance time covers the range 60–140 minutes (Dacie and Lewis, 1975). The measurement does not however differentiate between rapid clearance that is due to effective erythropoiesis and that due to ineffective erythropoiesis. For this differentiation it is necessary to measure the iron utilization or make a more sophisticated analysis of a clearance curve obtained over a longer period of time as discussed below.

Plasma iron turnover

The plasma iron turnover, which is the total amount of iron leaving the plasma in unit time, is calculated from the product of the fractional clearance rate of the iron with the total plasma iron as shown in equation (2). Here h is the clearance rate of iron/day and is equal to 0.693 divided by the $T_{1/2}$ clearance time measured in days; PI is the plasma iron concentration, PV is the plasma volume, and TBV is the total blood volume.

$$PIT = (h \times PI \times PV)/TBV \tag{2}$$

This expression is often approximated to the expression given in equation (3).

$$PIT = 1000\,PI\,(1 - PCV)/T_{1/2} \tag{3}$$

In both these equations the PIT is given in $\mu mol/\ell$ blood/day if the plasma iron is measured in $\mu mol/\ell$ and in equation (3) the $T_{1/2}$ is the half-clearance time in minutes. Unfortunately, however, the plasma iron turnover is of limited usefulness as there is again no distinction between effective and ineffective iron turnover and the results in health and disease overlap. The normal range is 72–144 $\mu mol/\ell$ of whole blood/day (Dacie and Lewis, 1975).

The use of equation (3) is less accurate than the use of equation (2) due to the assumed relationship between the plasma volume, the total blood volume and the PCV (as discussed in Chapter 3). In practice the plasma volume can be accurately measured using the iron-labelled transferrin and the red cell volume by using ^{99m}Tc-labelled red cells, as described in Chapter 3, so that accurate values of the plasma and total blood volumes can be simply obtained. The calculation of the plasma volume using the iron label is identical to that described in Chapter 3 using ^{125}I-labelled HSA except that the importance of extrapolating the plasma counts back to the time of injection is much greater with iron, as the loss of the label from the plasma pool is generally much faster than it is with HSA. It should be borne in mind that the measurement of the red cell volume is also required for the calculation of the iron utilization and will need to be measured at some stage of a ferrokinetic study in any case. *Table 5.1* illustrates the procedure for the calculation of the half-clearance time, plasma volume and plasma iron turnover.

Table 5.1 Calculation of plasma iron turnover

Weight of syringe containing labelled transferrin	= 23.246 g
Weight of syringe after removal of standard	= 22.636 g
Weight of standard	= 0.610 g
Weight of syringe after injection	= 17.055 g
Weight of injection	= 5.581 g
Ratio of injection to standard	= 9.15
Standard dilution volume	= 250 ml

Plasma clearance

	Sample counting								
			Time of sample (min)						
	Background	Standard	5	10	20	40	60	90	180
Volume (ml)	–	2.0	2.0	2.0	2.0	2.0	2.0	2.0	2.0
Counting time(s)	600	600	600	600	600	600	600	600	600
Counts	660	23811	18377	17341	16138	13623	11592	9127	4851
Counts/s	1.10	39.7	30.6	28.9	26.9	22.7	19.3	15.2	8.1
Corrected for background (counts/s)	–	38.6	29.5	27.8	25.8	21.6	18.2	14.1	7.0

Counts/s at time of injection (from logarithmic extrapolation) = 30.9
Half-clearance time = 80 minutes (from *Figure 5.1*)
Plasma volume = (39.7 × 250/2 × 9.15)/(30.9/2) = 2939 ml
Red cell volume (from study with labelled red cells) = 2215 ml
Plasma iron concentration (from chemical measurement) = 18 μmol/ℓ

Plasma iron turnover
$$= (0.693 \times 60 \times 24/80) \times 18 \times [2939/(2939 + 2215)]$$
$$= 128\,\mu\text{mol}/\ell/\text{day}$$

Iron utilization

For the measurement of the iron utilization by the red cells it is necessary to determine the amount of the injected iron that is incorporated into the circulating red cells. In order to do this, further blood samples are taken on alternate days for approximately two weeks after injection or until the amount of radioactive iron in the red cells reaches a plateau, as shown in *Figure 5.2*. The percentage iron utilization is given by the expression in

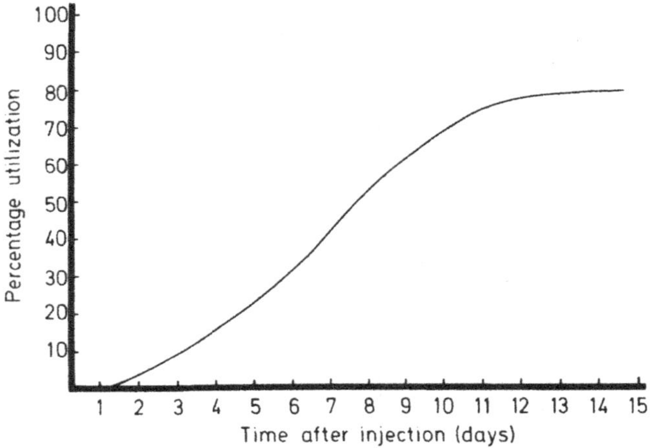

Figure 5.2 The utilization of ^{59}Fe by the red cells plotted as a function of time after the intravenous injection of ^{59}Fe labelled transferrin. The data are derived from the calculation shown in *Table 5.2*

equation (4) in which RCV is the red cell volume (measured in ml) and F is the ratio of the quantity of radioactivity/ml of red cells to the total radioactivity originally injected.

$$\text{percentage iron utilization} = \text{RCV} \times 100 \times F \qquad (4)$$

The red cell volume should be measured by one of the methods given in Chapter 3 if it has not already been measured for the PIT determination. In normal subjects the iron utilization rises steadily to a value of 70–80 per cent between days 10 and 14 of the study, whereas in aplastic anaemia it is typically 10–15 per cent and in ineffective erythropoiesis it is usually 30–50 per cent (Dacie and Lewis, 1975). However, if there is any significant haemolysis the

Table 5.2 Calculation of red cell utilization

Weight of plasma containing labelled transferrin injected	= 5.581 g
Weight of standard	= 0.610 g
Ratio of injection to standard	= 9.15
Standard dilution volume	= 250 ml

Sample counting

| | | | *Time of sample* (day) | | | | |
	Background	*Standard*	*3*	*6*	*10*	*12*	*14*
Volume (ml)	–	2.0	2.0	2.0	2.0	2.0	2.0
Counting times(s)	600	600	600	600	600	600	600
Counts	686	19049	2393	6662	13865	15292	15578
Counts/s	1.14	31.75	3.99	11.10	23.11	25.49	25.96
Corrected for background (counts/s)	–	30.61	2.85	9.96	21.97	24.35	24.82

Red cell volume (from study with labelled red cells) = 2215 ml

Percentage utilization on day 14
$$= 2215 \times 100 \times (24.82/2.0)/(30.61 \times 9.15 \times 250/2)$$
$$= 78.5 \text{ per cent (as shown in } Figure\ 5.2)$$

utilization will be underestimated due to the premature loss of the iron in the haemolysed red cells from the circulation. The calculation of the utilization from the raw data is illustrated in *Table 5.2.*

The use of physiological models and the calculation of the mean red cell life

The above measurements are very simple and straightforward but, as mentioned, they are not as clinically useful as could be desired nor is it possible to assess the relative amounts of effective and ineffective erythropoiesis accurately in a particular patient. Because of this, numerous attempts have been made to analyse the plasma clearance curve in more detail using compartmental models, and these have been reviewed by Najean *et al.* (1967) and Cavill and Ricketts (1974). The use of these models has been limited by the necessity to define and quantify physiological compartments that are not readily accessible to measurement, and none of these models has yet been universally accepted or adopted (Berlin, 1977). More recently a mathematical analysis of the plasma clearance curve has been developed which allows the calculation of various ferrokinetic parameters (Ricketts, Jacobs and Cavill, 1975; Cavill *et al.*, 1976; Ricketts *et al.*, 1977). Although this is not in very widespread use, it has been successfully used in several centres (Dinant and de Maat, 1978; Gupta *et al.*, 1979) and is briefly discussed here because it does enable a measurement of both effective and ineffective erythropoiesis to be made.

To use this analysis it is necessary to determine the plasma radioactivity for two weeks after the initial injection in order that an extended plasma clearance curve may be analysed. It is necessary to take extreme care with the blood samples because by the end of two weeks the radioactivity in the red cells is generally about 1000 times greater than it is in the plasma and any haemolysis that occurs in the sample can invalidate the results.

The protocol used by Cavill *et al.* (1976) for obtaining this extended plasma clearance curve is as follows. Venous blood is taken as 12 ml samples at approximately 10, 45 and 75 minutes, 4, 8 and 12 hours and 1, 3, 5, 7, 10 and 14 days after injection, and the exact time intervals are noted. After gentle separation of the plasma from the red cells, 4 ml samples of plasma are removed and the iron concentration of each sample is measured, so that corrections can be made for any variation in the plasma iron during the course of the study. (This correction is made by multiplying the measured count rate from each sample by the ratio of the initial iron concentration on the first day of the study to that of the sample.) In addition, any ^{59}Fe from the plasma sample that due to trace haemolysis is bound to haemoglobin rather than transferrin is removed from the samples before counting. This is achieved by adding an identical volume of a solution containing 30 ml of thioglycollic acid and 100 g of trichloracetic acid in 1 ℓ of 2M HCl to each plasma sample. This solution dissociates the iron from the transferrin, and leaves the denatured protein and the insoluble haem compounds to be precipitated by centrifugation. The supernatant then contains radioactivity originating from the transferrin only and equal volumes of supernatant are taken as samples to be counted. The 14-day plasma curve is fitted by computer to the triple exponential function given in equation (5) where $\alpha_1 + \alpha_2 + \alpha_3 = 1$ and f(t) is the fraction of the injected iron in the plasma at time t after injection.

$$f(t) = \alpha_1 e^{-\lambda_1 t} + \alpha_2 e^{-\lambda_2 t} + \alpha_3 e^{-\lambda_3 t} \tag{5}$$

From the computer fit of the experimental data to this function values are obtained for the quantities α_1, α_2, α_3, λ_1, λ_2 and λ_3 and these quantities together with the initial clearance rate $h (= \alpha_1\lambda_1 + \alpha_2\lambda_2 + \alpha_3\lambda_3)$ are used to calculate the four constants C_1, C_2, C_3 and C_4 which are given below in equations (6) to (9). Physiologically these constants relate to the refluxes of iron back into the plasma—the interested reader should refer to the original papers for more details (Cook *et al.*, 1970; Ricketts, Jacobs and Cavill, 1975).

$$C_1 = \tfrac{1}{2}\{\lambda_1 + \lambda_2 + \lambda_3 - h + [(\lambda_1 + \lambda_2 + \lambda_3 - h)^2 - 4(\alpha_1\lambda_2\lambda_3 + \alpha_2\lambda_3\lambda_1 + \alpha_3\lambda_1\lambda_2)]^{1/2}\} \tag{6}$$

$$C_2 = \tfrac{1}{2}\{\lambda_1 + \lambda_2 + \lambda_3 - h - [(\lambda_1 + \lambda_2 + \lambda_3 - h)^2 - 4(\alpha_1\lambda_2\lambda_3 + \alpha_2\lambda_3\lambda_1 + \alpha_3\lambda_1\lambda_2)]^{1/2}\} \tag{7}$$

$$C_3 = C_1^2/(C_1 - C_2) - C_1(\lambda_1\lambda_2 + \lambda_2\lambda_3 + \lambda_3\lambda_1 - C_1C_2 - \lambda_1\lambda_2\lambda_3C_1)/h(C_1 - C_2) \tag{8}$$

$$C_4 = C_1 + C_2 - C_3 - (\lambda_1\lambda_2 + \lambda_2\lambda_3 + \lambda_3\lambda_1 - C_1C_2)/h \tag{9}$$

The plasma iron turnover is calculated using equation (2) and a measurement of the plasma and red cell volumes, and the red cell iron utilization is calculated using equation (4) as described above. An alternative expression of the red cell iron utilization at time t after injection is, however, to write it as in equation (10) where k is the fraction of the PIT that is going to the formation of mature red cells, and the product of h and

$$\int_0^t f(t)\mathrm{d}t$$

is the total amount of iron that has passed through the plasma during this time.

$$\text{percentage iron utilization} = 100\,kh \int_0^t f(t)\mathrm{d}t \qquad (10)$$

Now $\displaystyle\int_0^t f(t)\mathrm{d}t,$

which is the area under the clearance curve up to time t after injection, can be accurately determined by computer and h, the initial clearance rate, may also be determined from the computed values of the α and λ constants (*see* above). Thus a value for k can be obtained from a combination of equations (4) and (10) from which k is given by the expression in equation (11).

$$k = \text{RCV} \times F/h \int_0^t f(t)\mathrm{d}t \qquad (11)$$

A measure of the amount of effective erythropoiesis occurring is given by the product of k with the plasma iron turnover and this has been termed the red cell iron turnover (RCIT) and is given by the expression in equation (12).

$$\text{RCIT} = k \times \text{PIT} \qquad (12)$$

A measure of the amount of ineffective erythropoiesis occurring depends on the refluxes; it is given by the expression in equation (13). This quantity has been termed the ineffective red cell iron turnover (IIT).

$$\text{IIT} = (C_4/C_2) \times \text{PIT} \qquad (13)$$

It is also possible to obtain a value for the mean red cell life from the measurement of red cell iron turnover, provided that the total red cell mass is in a steady state (ie provided that the daily loss of iron from the red cells due to red cell destruction is balanced by the production of new red cells containing the same quantity of iron). The mean red cell life is given by the product of the total red cell iron and the reciprocal of the red cell iron turnover, and may be calculated from the expression in equation (14), where Hb is the haemoglobin concentration in whole blood in g/dl, 62 µmol is the iron content of 1g of haemoglobin and the other quantities are as already defined above.

$$\text{Mean cell life} = \text{total red cell iron (µmol)/RCIT (µmol/day)} \quad (14)$$
$$= (\text{TBV} \times Hb \times 62)/(\text{RCIT} \times 100)$$

An alternative general method for obtaining ferrokinetic parameters and the mean red cell lifespan has been developed by Dagg *et al.* (1972). This is based on the use of the occupancy principle (Orr and Gillespie, 1968) which states that, in any compartmental system in a steady state, the ratio of the occupancy of a tracer in a compartment to the capacity of the compartment is a constant for all compartments in the system and is equal to the reciprocal of the flow of material through the system. Using this principle it has been shown that the mean red cell life can be determined from the expression in equation (15).

$$\text{MCL} = \frac{\text{occupancy of tracer iron in plasma} \times \text{stable iron in red cells}}{\text{stable iron in plasma} \times \text{iron utilization in red cells}} \quad (15)$$

The quantities of stable iron in the red cells and plasma are measured chemically (these are the capacities of the two compartments) and the iron utilization of the red cells (effectively the occupancy of the tracer in the red cells divided by the MCL) is determined in the normal way. The occupancy of the tracer in the plasma is obtained from a measurement of the area under the plasma clearance curve extrapolated to infinite time.

Both these methods provide results which, in general, correlate well with the usual methods of determining red cell lifespan (Hutcheon *et al.*, 1977; Ricketts, Cavill and Napier, 1977), but there is a discrepancy between the two methods with some patients (Orr *et al.*, 1979; Cavill and Ricketts, 1979; Bell and Orr, 1980). These methods both have the advantage that there are no elution errors to correct for as there are with ^{51}Cr (*see* Chapter 4) but on the debit side there is no survival curve that could be analysed for information on the destruction mechanism that occurs. However, if ferrokinetic studies are being made, and it is desired to measure the mean red cell life, the use of one of these techniques may well be preferred to making a separate measurement using ^{51}Cr or DFP labelled red cells. Alternatively, if an accurate elution correction is required for a ^{51}Cr study in order that the survival curve may be accurately plotted and analysed, one of these methods may be used to determine the mean red cell life at the same time as it is determined with ^{51}Cr so that the correct elution factor can be obtained and applied to the ^{51}Cr results.

References

BELL, R. and ORR, J. S. (1980) Ferrokinetic analysis of clearance curves. *British Journal of Haematology*, **45**, 165–166

BERLIN, N. I. (1977) The ICSH panel on diagnostic application of radioisotopes in haematology. *Clinics in Haematology*, **6**, 748–749

CAVILL, I. and RICKETTS, C. (1974) The kinetics of iron metabolism. *Iron in Biochemistry and Medicine*. pp. 613–647. Eds. Jacobs, A. and Worwood, M. London: Academic Press

CAVILL, I., RICKETTS, C., NAPIER, J. A. F., JACOBS, A., TREVETT, D. and BISHOP, R. D. (1976) The measurement of ^{59}Fe clearance from the plasma. *Scandinavian Journal of Haematology*, **17**, 160–166

CAVILL, I. and RICKETTS, C. (1979) Erythropoiesis and iron kinetics. *British Journal of Haematology*, **42**, 158

COOK, J. D., MARSAGLIA, G., ESCHBACK, J. W., FUNK, D. D. and FINCH, C. A. (1970) Ferrokinetics: a biologic model for plasma iron exchange in man. *Journal of Clinical Investigation*, **49**, 197–205

DACIE, J. V. and LEWIS, S. M. (1975) *Practical Haematology*, 5th Ed. pp. 445–450. Edinburgh and London: Churchill Livingstone

DAGG, J. H., HORTON, P. W., ORR, J. S. and SHIMMINS, J. (1972) A direct method of determining red cell lifespan using radioiron: an application of the occupancy principle. *British Journal of Haematology*, **22**, 9–19

DINANT, H. J. and DE MAAT, L. E. M. (1978) Erythropoiesis and mean red cell lifespan in normal subjects and patients with the anaemia of active rheumatoid arthritis. *British Journal of Haematology*, **39**, 437–444

GUPTA, M. M., ROTH, P., WERNER, E. and KALTWASSER, J. P. (1979) Application of ferrokinetic investigation for differential diagnosis in bone marrow hypoplasia and their clinical relevance. *European Journal of Nuclear Medicine*, **4**, 17–20

HAHN, P. F., BALE, W. F., LAURENCE, E. O. and WHIPPLE, G. H. (1939) Radioactive iron and its metabolism in anaemia. *Journal of Experimental Medicine*, **69**, 739

HUTCHEON, A. W., HORTON, P. W., ORR, J. S. and DAGG, J. H. (1977) The assessment of red cell survival in normal subjects and in patients with haemolytic disorders and ineffective erythropoiesis using the radioiron occupancy method. *British Journal of Haematology*, **37**, 195–205

ICSH (1978) The measurement of total and unsaturated iron binding capacity in serum. *British Journal of Haematology*, **38**, 281–290

ICSH (1978) Recommendations for measurement of serum iron in human blood. *British Journal of Haematology*, **38**, 291–294

NAJEAN, Y., DRESCH, C., ARDAILLOU, N. and BERNARD, J. (1967) Iron metabolism – study of different kinetic models in normal conditions. *American Journal of Physiology*, **213**, 533–546

ORR, J. S. and GILLESPIE, F. C. (1968) Occupancy principle for radioactive tracers in steady state biological systems. *Science*, **162**, 138–139

ORR, J. S., HORTON, P. W., HUTCHEON, A. W. and DAGG, J. H. (1979) Erythropoiesis and iron kinetics. *British Journal of Haematology*, **42**, 155–157

RICKETTS, C., JACOBS, A. and CAVILL, I. (1975) Ferrokinetics and erythropoiesis in man: the measurement of effective erythropoiesis ineffective erythropoiesis and red cell lifespan using ^{59}Fe. *British Journal of Haematology*, **31**, 65–75

RICKETTS, C., CAVILL, I., NAPIER, J. A. F. and JACOBS, A. (1977) Ferrokinetics and erythropoiesis in man: an evaluation of ferrokinetic measurements. *British Journal of Haematology*, **35**, 35–41

RICKETTS, C., CAVILL, I. and NAPIER, J. F. (1977) The measurement of red cell lifespan using ^{59}Fe. *British Journal of Haematology*, **37**, 403–408

6

External counting

In both the cell-survival and the ferrokinetic studies discussed in the last two chapters, additional clinically useful information may be obtained by measuring the quantity of the radionuclide tracer that is taken up in the various organs. In the case of red cell survival measurements it is of value to know the amounts of red cell destruction that occur in the liver and spleen, rather than in the body as a whole, and this is especially the case if a splenectomy is being considered. Similarly, in ferrokinetic studies, it is of value to know the amounts of erythropoiesis that occur in different parts of the body.

Until relatively recently these measurements were always made with an external uptake counter placed over the organ of interest and it was only possible to obtain qualitative information. Over the last 10 years or so, however, methods have been developed which allow quantitative measurements to be made using radionuclide imaging equipment and these measurements can also be used to calibrate the simple external detectors. In this chapter the use of the simple technique is described, and the quantitative technique and its other applications in haematology are discussed in the next chapter.

External counting equipment

The simplest equipment that can be used for external counting consists of a single lead-shielded scintillation detector (at least 75 mm in diameter and 37 mm thick) with a single hole collimator, as described in Chapter 2. The dimensions recommended by the International Committee for Standardization in Haematology for the collimator hole are 50 mm in diameter and 70 mm deep (ICSH, 1975). However, it has been shown that superior results can be obtained if two detectors are used (Bowring and Glass, 1974) and also if more sophisticated multiple hole collimators are used (Bowring and Glass, 1974; Clarke, Duffy and Malone, 1978). The use of two detectors, positioned opposing each other with the patient in between, results in better uptake curves and in a counting system where the sensitivity of the equipment is largely independent of the position of the organ between the detectors. The use of multiple holes in the collimator results in a sharper edge to the field of view of the detectors (ie a smaller penumbra) and this ensures that the detection of γ-rays from tissue other than that immediately between the detectors is reduced to the minimum. A diagram of a seven hole collimator which has been found satisfactory for both ^{51}Cr and ^{59}Fe measurements is shown in *Figure 6.1*. *Figure 6.2* shows the detectors of a typical dual detector system in which the patient under investigation lies on a couch between the two detectors.

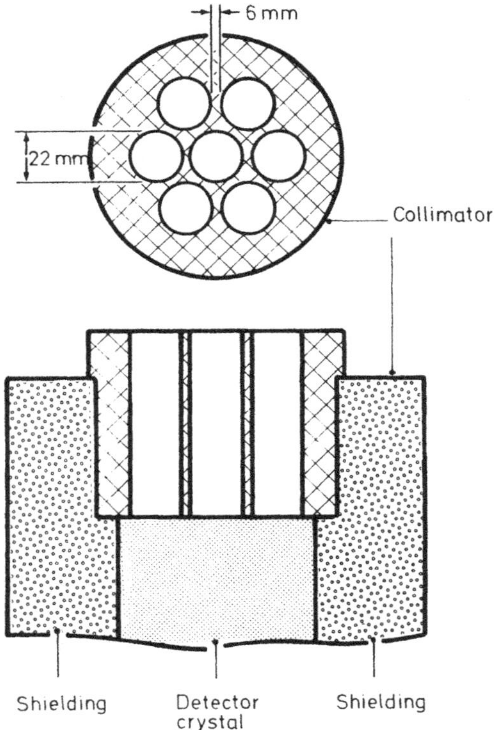

Figure 6.1 A seven-hole collimator suitable for external counting

Figure 6.2 The two detectors of a dual detector external counting system. (From Bowring and Glass (1974) by courtesy of the authors, editors and publishers.)

Procedure for ^{51}Cr labelled red cell studies

When making measurements over the sites of red cell destruction in conjunction with a red cell survival study, the measurements are started 30–60 minutes after injection of the ^{51}Cr labelled cells. A 60 minute delay should be used for patients with splenomegaly in whom the mixing time may be prolonged. The measurements are then repeated daily or on alternate days, depending on the rate of haemolysis, for 1–2 weeks. It is very important that great care is taken to position the detector/s in exactly the same way each day and also to ensure that the patient is always counted in the same posture. If possible, each set of measurements should be done at the same time each day. The ICSH has recommended two methods for positioning the detectors, one for single detector systems and one for dual detector systems (ICSH, 1975).

Single detector positioning

For all measurements the patient lies supine and the detector is positioned so that the collimator is just touching the skin. For the heart measurement the detector is vertical and is centred over the mid-line at the level of the third interspace. (A single detector usually produces more counts in this position over the great vessels than directly over the heart.) For the liver measurement the detector is vertical and is positioned half-way between the right midclavicular and anterior axillary lines at a distance of 30–40 mm above the costal margin. For the spleen the detector is placed horizontally over the position found to give the highest count rate. Some workers have recommended placing the patient in the right lateral position for spleen counting and keeping the detector vertical, but this position is more difficult to reproduce accurately and is often uncomfortable for the patient.

Dual detector positioning

For all measurements the patient lies supine and the detectors are vertical. The separation of the detectors for each organ must be noted when the patient is first counted and the same separation used on subsequent occasions. For the heart measurement the upper detector is aligned over the third interspace at the left sternal border, for the liver it is positioned in the same way as for a single detector and for the spleen the position giving the highest count rate is selected. An alternative method of determining the splenic counting position is to localize the organ by prior imaging using ^{99m}Tc-labelled heat-damaged red cells (*see* Chapter 7) and then to mark a position on the skin over the centre of the organ.

Counting procedure and analysis of the data

On each day that counting is performed a standard must be counted under identical conditions in order that the measured count rate from it may be used for correction for radioactive decay and for variations in the sensitivity of the detectors and ancillary equipment. A source of about 50 kBq is suitable for this and for dual detector systems should be counted in a special jig which positions it midway between the two detectors. Any gross change in the standard count from that expected should be investigated before the patient is counted. The background count rate of the detectors must also be measured each day.

The detector count rate should be determined over each organ in turn and these measurements should, if possible, all be repeated a second time to check the consistency of the detector positioning. If one of the repeat counts differs by more than about 5 per cent from the first count the measurement should be repeated. At least 2500 counts should be recorded over each organ in order to determine the count rate. The equipment can usually be set so that either the time taken to record this number of counts is measured or the counting time is preset to a value that ensures detection of at least this number of counts in the counting time. Several methods have been used in the past for presenting the results of ^{51}Cr external counting (eg Jandl *et al.*, 1956; Hughes-Jones and Szur, 1957; Lewis, Szur and Dacie, 1960) and the method now recommended by the ICSH (1975) is based on that of Lewis, Szur and Dacie. The principle of this method is that after the initial measurements on the first day of the study, the heart count rate is used to predict the count rates that would be measured over the spleen and liver if the initial count rate obtained over each of these organs fell at the same rate as over the heart (ie the count rate expected if there were no destruction of cells nor accumulation of ^{51}Cr in the organs). The difference between the observed count rate and the predicted count rate is termed the excess count and is taken as a measure of the accumulation of ^{51}Cr in the organ. The count rates obtained over each organ are corrected for background and then for decay and detector sensitivity using the measurements of the standard sample. In dual detector systems the counts recorded by each detector must first be summed, although this is often done electronically in the equipment itself. The count rate over the heart on the first day (day 0 of the study) is designated as 1000 and the normalizing factor required for this is then used to normalize the count rates over the other organs to this figure. The count rate over the heart falls at the same rate as the level of radioactivity in the blood (Ferrant and Bowring, 1974) except in very rare cases in which there is an accumulation of ^{51}Cr in cardiac or lung tissue that causes the count rate to fall more slowly or even to rise. In such cases, after the initial heart count, it is necessary to use the blood data instead of the heart counts to calculate the expected count rates over the other organs. This, in fact, is a more accurate procedure as there is usually less scatter of the points around the curve if blood data are used (Najean *et al.*, 1975).

When the normalized count has been obtained it is compared with the predicted count and the difference, the excess count, is calculated. Finally the ratio of the observed count rate over the spleen to that over the liver is calculated both in absolute terms and in terms normalized to a value of 1.0 for the first day of the study. This is intended to show the relative amounts of destruction that occur in the spleen and liver. *Table 6.1* shows a typical set of

Table 6.1 Calculation of the results from ^{51}Cr external counting data

Day of study	0	1	3	5	7	9	11	14
Measured count rate (counts/s)								
Background	3.3	3.4	3.4	3.5	3.1	3.3	3.5	3.4
Standard	38.0	36.5	35.3	33.9	32.3	30.9	29.9	27.7
Corrected for background	34.7	33.1	31.9	30.4	29.2	27.6	26.4	24.3
Standard correction factor	1.00	1.05	1.09	1.14	1.19	1.26	1.31	1.43
Heart	48.3	42.8	34.8	30.3	27.0	24.0	21.4	19.5
Corrected for background	45.0	39.4	31.4	26.8	23.9	20.7	17.9	16.1
Corrected for standard	45.0	41.4	34.2	30.6	28.4	26.1	23.4	23.0
Spleen	46.7	47.1	48.6	48.5	47.7	46.7	45.8	42.6
Corrected for background	43.4	43.7	45.2	45.0	44.6	43.4	42.3	39.2
Corrected for standard	43.4	45.9	49.3	51.3	53.1	54.7	55.4	56.1
Liver	33.5	30.6	27.8	25.0	23.5	21.5	20.4	18.9
Corrected for background	30.2	27.2	24.4	21.5	20.4	18.2	16.9	15.5
Corrected for standard	30.2	28.6	26.6	24.5	24.3	22.9	22.1	22.2
Normalized counts								
Heart	1000	920	760	680	631	580	520	511
Spleen	964	1020	1096	1140	1180	1216	1231	1246
Predicted	964	887	733	656	608	559	501	493
Excess	–	133	363	484	572	657	730	753
Liver	671	636	591	544	540	509	491	493
Predicted	671	585	510	456	423	389	349	343
Excess	–	51	81	88	117	120	142	150
Spleen/liver	1.44	1.60	1.85	2.10	2.19	2.39	2.51	2.53
Normalized spleen/liver	1.00	1.11	1.28	1.46	1.52	1.66	1.74	1.76

data and the manipulations required to obtain the excess counts and spleen:liver ratios. Finally, the data are plotted on ordinary graph paper as shown in *Figure 6.3*.

One of four general patterns is usually observed from these measurements (Dacie and Lewis, 1975), as shown in *Figure 6.4*.

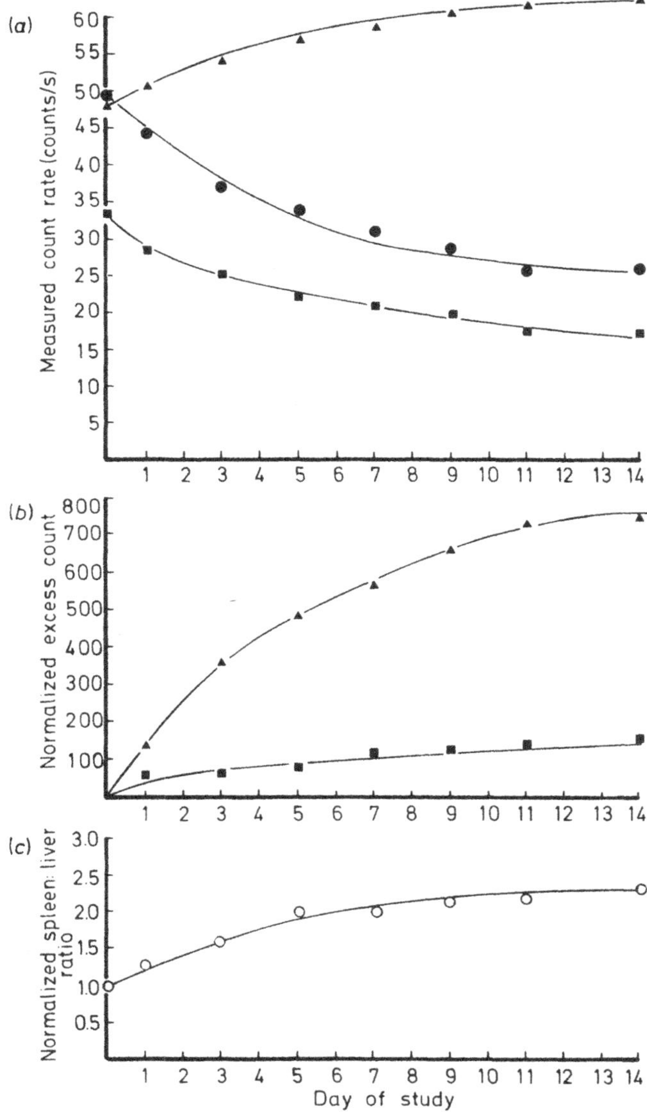

Figure 6.3 The final data from *Table 6.1* plotted. (*a*) measured count rates. (*b*) excess counts. (*c*) normalized spleen/liver ratio. Spleen readings are designated by ▲, heart readings by ● and liver readings by ■

1. Little or no excess accumulation in either organ;
2. Excess accumulation of ^{51}Cr in the spleen only;
3. Excess accumulation in the liver only;
4. Excess accumulation in both organs.

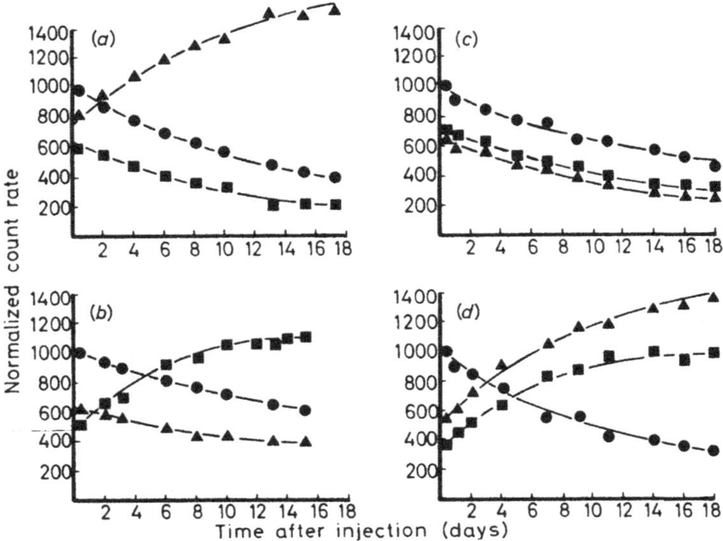

Figure 6.4 Typical ^{51}Cr external counting patterns showing the normalized external counting patterns obtained in four patients. (*a*) excess counts observed over the spleen. (*b*) excess counts observed over the liver. (*c*) no excess counts over either spleen or liver. (*d*) excess counts over both the liver and spleen. Spleen readings are designated by ▲, heart readings by ● and liver readings by ■

Splenectomy is usually of benefit to patients showing excess accumulation in the spleen alone and to a lesser extent excess accumulation in both the spleen and the liver. However, the degree of benefit does not correlate well with the magnitude of the excess count (Ahuja, Lewis and Szur, 1972) and any decision on the advisability of performing a splenectomy must take account of all other laboratory data and the clinical features of the disease. Among the variables that contribute to this poor correlation are first the volume of the organ in the field of view of the detector/s in relation to the amount of other tissue in the field of view and to the total volume of the organ; secondly, the thickness of the tissue surrounding the organ and, thirdly, the loss of ^{51}Cr from the organ after cells have been destroyed in it. The next chapter discusses how these sources of error can be reduced and the whole technique made more reliable.

Procedure for ^{51}Cr labelled platelet studies

Although external counting with ^{51}Cr labelled platelets is not an investigation that is as universally performed as the investigation with red cells, some centres consider it to be a useful test. The techniques used are the same as those used for red cells and the same organs are counted. The measurements need to be made at more frequent intervals, however, because of the shorter lifespan of platelets. The results suffer from the same inaccuracies as the simple red cell studies and, in addition, the interpretation of the results is made more difficult by the role of the spleen as a pool for platelets. In normal subjects the spleen can pool up to 30 per cent of the platelet mass and in splenomegaly this figure can rise as high as 90 per cent. These factors make the test of questionable value in determining whether or not a patient might benefit from splenectomy (Najean, Dresch and Bernard, 1967; Aster and Keen, 1969; Dacie and Lewis, 1975), and quantitative imaging of platelets labelled with ^{111}In (*see* Chapter 7) may be expected to become the preferred method for investigating the *in vivo* distribution of platelets.

Procedure for ferrokinetic studies

For ferrokinetic studies two sets of measurements are made. First, measurements are made over the first few hours after the injection of radioactive iron while the iron is being cleared from the plasma, and secondly, measurements are made over the following days to study the release of the utilized iron back into the blood. The detector positioning is identical to that described for ^{51}Cr with the exception that, in addition to counting over the heart, spleen and liver, the detector is placed vertically over the sacrum (with the patient prone if a single detector is used) so that a typical region of bone marrow may be counted.

The first measurements are commenced approximately 5 minutes after injection and then repeated at 20, 40, 60, 90, 120, 180, 240 and 300 minutes. The second set of measurements should be made using the same counting positions. These measurements are started the following day and then repeated on alternate days for the next 10–14 days. The initial count rate over each organ after injection is generally expressed as 100 and subsequent measurements (after the necessary corrections for background, decay of the radionuclide and detector sensitivity have been made) are expressed as a percentage of this. Normally ^{59}Fe is used for this study because it is preferred for the other ferrokinetic studies that will be simultaneously carried out. For the first day's data, however, [52]Fe can be used if it has been administered to obtain

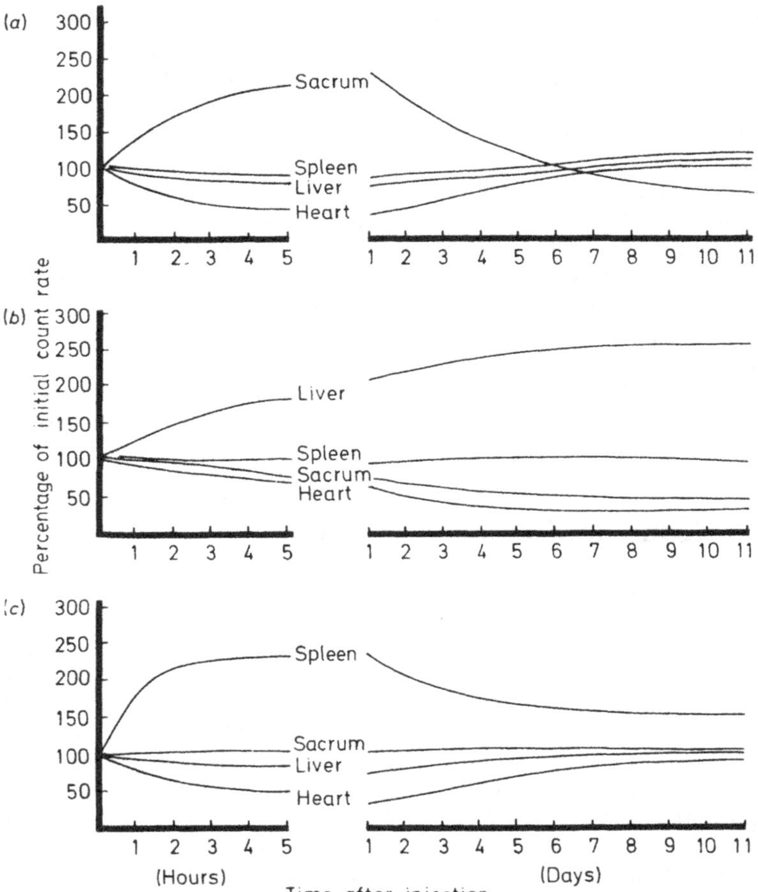

Figure 6.5 Typical ^{59}Fe external counting patterns. (*a*) normal pattern. (*b*) pattern from a patient with aplastic anaemia. (*c*) pattern from a patient with myelosclerosis

images of the distribution of functioning bone marrow (*see* Chapter 7).

Probably the greatest use of external counting with ^{59}Fe is in patients with extramedullary erythropoiesis when it is desired to know the sites and extent of this condition. Typical results observed (*a*) in haematologically normal patients, (*b*) in aplastic anaemia and (*c*) with myelosclerosis are shown in *Figure 6.5*. Note the large increase and fall in the count rate over the sacrum and the opposite effect over the heart in the normal patient, the steady rise in count rate over the liver in aplastic anaemia showing iron storage but not erythropoiesis, and the increase and fall in the spleen count in the case of myelosclerosis which demonstrates some effective extramedullary erythropoiesis in this organ.

As with the other simple external counting investigations it must be stressed that these results can only be qualitative. Even so, they are usually a useful addition to the ferrokinetic measurements discussed in the previous chapter. A technique for quantifying the role of the spleen is described in the next chapter.

Simultaneous external counting with ^{51}Cr and ^{59}Fe

When it is necessary to perform both ^{51}Cr and ^{59}Fe studies on a patient, the greatest precision is achieved by performing ferrokinetic external counting studies after ^{51}Cr studies have been completed because there is then no need to correct the ^{51}Cr results for the presence of ^{59}Fe. However, this is a very time-consuming prodecure and effectively doubles the length of time required for the studies, and it is possible to perform both sets of measurements virtually simultaneously. For this combined study it is necessary to start the ^{59}Fe external counting at least 24 hours before the ^{51}Cr labelled cells are injected. On the day when the ^{51}Cr study is to start each organ of the patient must be counted before the injection of the ^{51}Cr labelled cells, with the pulse height analysers of the counting equipment set both for ^{59}Fe and for ^{51}Cr. This is necessary so that the fraction of the count rate in the ^{59}Fe window that appears in the lower energy ^{51}Cr window due to the detection of scattered ^{59}Fe rays may be calculated for each organ. It is then assumed that this fraction will remain the same for each organ for the rest of the study and the subsequent counts obtained in the ^{51}Cr window are appropriately corrected. Results from the first two days of a combined study are shown in *Table 6.2* to illustrate the calculation procedure.

Table 6.2 Calculation of the data from the first 2 days of a combined study with ^{51}Cr and ^{59}Fe. The figures given are measured count rates (counts/s)

Day of study	Before ^{51}Cr injection		After ^{51}Cr injection			
	0	1	0	1	1	2
Counting channel	Cr	Fe	Cr	Fe	Cr	Fe
Background	3.6	4.2	3.6	4.2	3.5	4.3
^{59}Fe standard	–	52.6	–	52.6	–	51.8
Corrected for background	–	48.4	–	48.4	–	47.5
Standard correction factor	–	1.01*	–	1.01	–	1.03
^{51}Cr standard	–	–	74.5	–	70.4	–
Corrected for background	–	–	70.9	–	66.9	–
Standard correction factor	–	–	1.00	–	1.06	–
Heart	5.8	19.4	25.3	19.4	21.9	20.2
Corrected for background	2.2	15.2	21.7	15.2	18.4	15.9
Corrected for cross-talk	–	–	19.5	–	16.1	–
Corrected for standard	–	15.4	19.5	15.4	17.1	16.4
Spleen	6.5	22.5	24.4	22.5	27.6	22.1
Corrected for background	2.9	18.3	20.8	18.3	24.1	17.8
Corrected for cross-talk	–	–	17.9	–	21.3	–
Corrected for standard	–	18.5	17.9	18.5	22.6	18.3
Liver	6.8	25.6	19.4	25.6	19.4	25.0
Corrected for background	3.2	21.4	15.8	21.4	15.9	20.7
Corrected for cross-talk	–	–	12.6	–	12.8	–
Corrected for standard	–	21.6	12.6	21.6	13.6	21.3
Sacrum	–	30.4	–	30.4	–	27.2
Corrected for background	–	26.2	–	26.2	–	22.9
Corrected for standard	–	26.5	–	26.5	–	23.6

^{51}Cr cross-talk correction is derived thus: heart, Cr $-$ (Fe $\times$ 2.2/15.2); spleen, Cr $-$ (Fe $\times$ 2.9/18.3); liver, Cr $-$ (Fe $\times$ 3.2/21.4).

* The standard correction factor for ^{59}Fe on day 1 is 1.01 rather than 1.00 because of decay since day 0 of the ^{59}Fe study. The corrected standard count rate on day 0 was 48.9

References

AHUJA, S., LEWIS, S. M. and SZUR, L. (1972) Value of surface counting in predicting response to splenectomy in haemolytic anaemia. *Journal of Clinical Pathology*, **25**, 467–472

ASTER, R. H. and KEENE, W. R. (1969) Sites of platelet destruction in idiopathic thrombocytopenic purpura. *British Journal of Haematology*, **16**, 61–73

BOWRING, C. S. and GLASS, H. I. (1974) Collimation and surface counting in haematology. *Journal of Clinical Pathology*, **27**, 751–756

CLARKE, L. P., DUFFY, G. J. and MALONE, J. F. (1978) An improved uptake probe designed for a large crystal rectilinear scanner. *Physics in Medicine and Biology*, **23**, 118–126

DACIE, J. V. and LEWIS, S. M. (1975) *Practical Haematology*, 5th Edn. p. 390. Edinburgh and London: Churchill Livingstone

FERRANT, A. and BOWRING, C. S. (1974) *Unpublished data*

HUGHES-JONES, N. C. and SZUR, L. (1957) Determination of the sites of red cell destruction using ^{51}Cr labelled cells. *British Journal of Haematology*, **3**, 320

ICSH (1975) Recommended methods for surface counting to determine sites of red cell destruction. *British Journal of Haematology*, **30**, 249–254

JANDL, J. H., GREENBURGH, M. S., YONEMOTO, R. H. and CASTLE, W. B. (1956) Clinical determination of the sites of red cell sequestration in haemolytic anaemias. *Journal of Clinical Investigation*, **35**, 842

LEWIS, S. M., SZUR, L. and DACIE, J. V. (1960) The pattern of erythrocyte destruction in haemolytic anaemia as studied with radioactive chromium. *British Journal of Haematology*, **6**, 122

NAJEAN, Y., DRESCH, C. and BERNARD, J. (1967) The platelet destruction site in thrombocytopenic purpuras. *British Journal of Haematology*, **13**, 409–426

NAJEAN, Y., CACCHIONE, R., DRESCH, C. and RAIN, J. D. (1975) Methods of evaluating the sequestration site of red cells labelled with ^{51}Cr: a review of 96 cases. *British Journal of Haematology*, **29**, 495–510

7

Imaging and applications of quantitative imaging

The basic design of radionuclide imaging equipment is discussed in Chapter 2. As is noted there, unless special care is taken to calibrate the equipment, the images obtained give only qualitative rather than quantitative informaton about the distribution of the radionuclide. While the qualitative image is of value to the haematologist, the additional information obtainable from quantitative images is often useful. In this chapter, conventional qualitative imaging of the spleen and bone marrow are briefly considered, and then the calibration of imaging equipment and applications of quantitative imaging in haematology are discussed in detail.

Spleen imaging

The accumulation of suitable radionuclides in the spleen provides one of the simplest and safest means of imaging the spleen in order to demonstrate its size and position. The most widely used tracer for this is ^{99m}Tc-labelled heat-damaged red cells, although ^{51}Cr-labelled heat-damaged red cells and cells damaged with ^{197}Hg-labelled mercurihydroxypropane (MHP) have been used in the past. In all these cases the damaging procedure alters the rigidity of the surface of the red cells and causes sphering, which prevents them from passing through the microcirculation of the spleen and traps them within it. Colloids labelled with ^{99m}Tc and ^{113m}In are also commonly used for imaging both the spleen and the liver, but if an image of the spleen is the primary requirement ^{99m}Tc-labelled heat-damaged red cells are superior. This is because heat-damaged red cells are not taken up by the liver and therefore do not interfere with the image of the spleen. The uptake in the liver is often 10–20 times greater than the uptake in the spleen when colloids are used. The procedures for damaging and labelling a sample of the patient's red cells are described in Appendix 1.

The cells are normally labelled with 50–100 MBq of ^{99m}Tc and imaging can be commenced 15 minutes after injection. It is more usual, however, to wait about one hour so that the background blood radioactivity may reduce. This is especially important in patients with poor splenic function with whom it may be preferable to wait even longer. Anterior, posterior and left lateral images are usually obtained.

Splenic imaging is of use in demonstrating enlargement of the spleen, the position or abnormal position of the spleen, the presence of accessory splenic tissue and the presence of space-occupying lesions in the organ. It is also used to help identify the nature of masses in the left hypochrondrium and to assess splenic function when splenic atrophy is suspected (Pettit, 1977; Ramchandran, Margouleff and Atkins, 1980). Aspects of quantitative spleen imaging are discussed below.

Bone marrow imaging

The standard ferrokinetic measurements discussed in Chapters 5 and 6 provide only limited information about the sites of functioning haemopoietic marrow, and imaging of the marrow itself can be of considerable additional value. A variety of radionuclides are taken up by the cells in the bone marrow but of these the most commonly used have been ^{99m}Tc-labelled colloids, ^{111}In and ^{52}Fe. The use of colloid has never been popular as it shows only the distribution of phagocytic activity and, since over 95 per cent of the injected radioactivity is normally taken up in the liver and spleen, satisfactory images of the bone marrow are obtained only when the function of these organs is reduced from normal. Although in normal marrow the distribution of phagocytic and haemopoietic cells is the same, this is often not the case in diseases of the marrow (Van Dyke et al., 1967; Merrick et al., 1975).

Ionic ^{111}In and transferrin labelled with ^{111}In have also been used for bone marrow imaging because the radiation characteristics of ^{111}In make it technically better for imaging than any of the radioisotopes of iron and there is also some evidence that indium behaves physiologically as an analogue of iron (Lilien et al., 1973; Rayadu et al., 1973). However, here again it has been found (McIntyre et al., 1974; Merrick et al., 1975; Parmentier et al., 1977; Perry and Holmes, 1980) that there are marked differences between the metabolic behaviour of the two elements and that these are particularly noticeable in marrow disease. In red cell aplasia, for instance, ^{111}In gives no useful information about erythropoietic tissue distribution and in myelofibrosis the information can be misleading.

There is thus, at present, no alternative to using either ^{59}Fe or ^{52}Fe for obtaining images of the distribution of functioning haemopoietic cells in the reticuloendothelial system. Although images have been obtained using ^{59}Fe (Ronai *et al.*, 1969; Chaudhuri *et al.*, 1974; Parmentier *et al.*, 1977), these are not of high quality because of the small quantities of this radionuclide that can be administered and the very high energy of the γ-rays emitted. In addition, the images can be obtained only after modification to standard imaging equipment to allow for the high energy of the ^{59}Fe γ-rays.

Therefore, at present ^{52}Fe is the best radionuclide for obtaining these images but even this has several disadvantages:

1. It is not widely available as it can only be produced in a cyclotron and, in the UK, the sole source is the cyclotron run by the Medical Research Council at Hammersmith Hospital.
2. The amount that can be safely administered is limited by the presence of ^{55}Fe contamination (Thakur, Nunn and Waters, 1971).
3. Its γ-ray energy, although approximately half that of ^{59}Fe, is still too high for obtaining the best quality images with most imaging equipment.
4. Its 8 hour half-life means that only the early distribution of the iron can be visualized.

For these reasons the imaging of bone marrow is not as frequently performed as it would be if there was a satisfactory and more widely available radionuclide.

The use of ^{52}Fe has been found of value in the assessment of the distribution of functioning haemopoietic tissue, and especially for the selection of sites for biopsy in aplastic anaemia and myeloproliferative disorders, in both of which the distribution of functioning marrow can be very patchy (Merrick *et al.*, 1975). ^{52}Fe has also proved valuable for following the effects of various treatment regimens on erythropoiesis (Knospe *et al.*, 1976; Pettit, 1977; Pettit, Lewis and Nicholas, 1979). Images of the iron distribution are normally obtained 3–4 hours after intravenous injection of transferrin labelled with up to 10MBq of ^{52}Fe. Quantitative imaging of the spleen with ^{52}Fe for the investigation of extramedullary erythropoiesis in this organ is discussed below.

Principles of quantitative imaging

The primary reason why the images obtained from normal radionuclide imaging devices are not easily quantified is the change in sensitivity of the instrument to sources of γ-rays at different positions within the patient. In all scanners except the computerized tomographic scanners the further a source is from the detector the less of the radionuclide it will appear to contain in the resulting image. With the rectilinear scanner and the hybrid scanning systems mentioned in Chapter 2 this problem can be overcome by the use of two detectors with suitable collimators one above and one below the patient (Arimizu and Morris, 1969). The response of the system can then be made constant to within 5–10 per cent and independent of the position of the source between the detectors. It is more difficult with gamma cameras to make the response independent of source distance because of the much greater loss of sensitivity with increasing distance from the collimator, but correction is nevertheless possible (Ferrant and Cauwe, 1979; Fleming, 1979). Apart from the problem of the distance of the organ from the detector there are two further factors which need consideration. First, the number of detected γ-rays forming the image depends on the amount of tissue surrounding the organ since this governs the total number of γ-rays that are absorbed or scattered in the patient and which therefore do not contribute to the image. Secondly, the number of detected γ-rays depends on the total imaging time (or the speed of scanning) since this governs the effective time spent detecting γ-rays from the organ. Generally, the imaging time or scanning speed can be simply measured and this correction does not present any problem. A profile of the patient may also be easily obtained (for instance by bending a thin strip of lead round the patient), and this can then be used to provide measurements of the thickness of the patient so that correction may be made for the absorption and attenuation of the γ-rays from the organ. When these corrections have been made, the 'corrected' number of γ-rays making up the image is, to a good approximation, proportional to the quantity of the radionuclide in the organ. However, if some of the radionuclide is still circulating in the blood or is taken up in the surrounding tissues it is necessary to make a correction for this as well.

For any radionuclide, therefore, it is possible, in principle, to calibrate the imaging equipment so that the quantity of the radionuclide in the organ of interest can be determined. The relationship between the quantity of the radionuclide and the number of γ-rays making up the corrected image of the organ will

differ however for different radionuclides. It is, therefore, necessary to calibrate the imaging equipment for each radionuclide with
which it is desired to obtain quantitative images. Since the imaging
equipment is usually to be found in hospital departments of
Medical Physics, Nuclear Medicine or Diagnostic Radiology
rather than Haematology, and because a physicist will probably be
needed to calibrate it, the details of the calibration and quantitative imaging techniques are described separately in Appendix 3.

Applications of quantitative imaging in haematology

The current clinical uses of quantitative imaging in haematology
cover three fields. The technique has been used to measure the
size of the pool of red cells in the spleen, it has been used to
calibrate ^{51}Cr external counting equipment, and it has been used
to measure the uptake of iron and to calibrate ferrokinetic external
detectors.

In addition, the technique has been used to study the distribution of ^{111}In labelled platelets and white cells (Lavender *et al.*,
1977) and it may be expected that when these techniques become
universally available this will be a field in which quantitative
imaging will be of great use to the haematologist. At present,
however, there is concern that the labelling process may damage
the cells which will then not behave in a truly physiological way
(Segal, 1978; Danpure, Osman and Hesselwood, 1979), and
alternative labelling techniques are under development (Sinn and
Silvester, 1979). However, although some difficulties have been
reported with the interpretation of the images obtained (Coleman
and Welch, 1980) the technique does provide a useful method for
localizing sites of infection by using white cells (Peters and
Lavender, 1980) and sites of thrombosis by using platelets (Davis
et al., 1978) and the centres developing the techniques are very
optimistic about their future applications.

Measurement of the splenic red cell pool

An enlarged spleen can contribute to an anaemia purely as a result
of the red cell pooling that occurs in the organ itself, However, the
actual amount of blood in the pool is not only proportional to the
size of the organ, but also depends on its internal structure. Thus it
has been shown that patients with primary polycythaemia have a

larger splenic pool per unit splenic mass than those with secondary polycythaemia (Bateman *et al.*, 1978). Similarly, the red cell pool has been studied in patients with myelosclerosis, malignant lymphomas and leukaemic reticuloendotheliosis, and differences in the ratio between the volume of the blood pool in the organ and the mass of the organ have been found (Pettit, *et al.*, 1971; Lewis *et al.*, 1977).

Previous methods proposed for measuring the volume of the splenic red cell pool have either relied on the slower mixing of labelled cells with the pool compared with their mixing with the rest of the body (Toghill, 1964; Pryor, 1967) or have required additional ultrasonic investigations (Christensen, 1971). At present quantitative imaging alone provides a direct measurement of the pool size and this measurement is a direct application of the quantitative imaging technique. Originally ^{11}C carbon monoxide was used as the red cell label (Glass *et al.*, 1968) but ^{99m}Tc is now used almost exclusively (Hegde *et al.*, 1973). A sample of the patient's red cells is labelled with about 100 MBq of ^{99m}Tc using the method described in Appendix 1 and the quantitative image is usually obtained about half an hour after reinjection of the cells, although it may be done later on patients that have spleens with slow mixing pools. The result may be expressed either as a percentage of the total red cell volume or, if the labelled cells are also used to measure the total red cell volume (*see* Chapter 3), as a volume in millilitres. Because of the elution of ^{99m}Tc from red cells

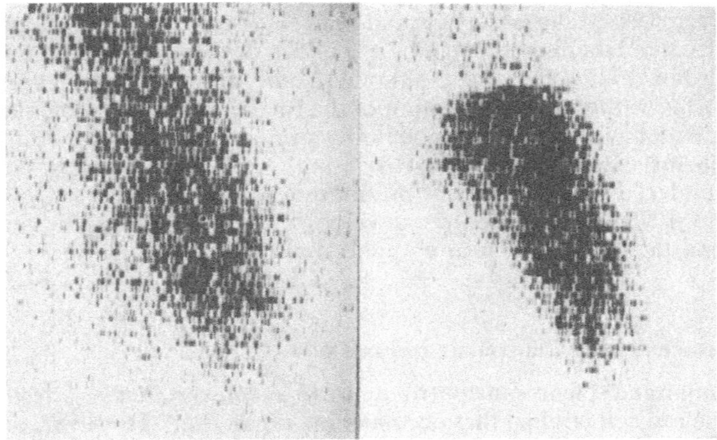

Figure 7.1 Splenic images obtained with a rectilinear scanner; left, with undamaged labelled cells and right, with heat-damaged labelled cells

(*see* Appendix 1), which is of the order of 5 per cent/hour, the measured percentage uptake will be reduced by this amount and a correction may be needed. However, provided the quantitative image is not taken later than one hour after injection, correction for elution of the label is not usually considered to be justified.

In some patients it is difficult to separate the images of the blood pools of the heart and spleen and this makes it difficult to decide on the limits of the organ when the uptake is calculated (*see* Appendix 3). An additional image of the spleen is therefore done with heat-damaged cells (*see* p.89) in order to define its boundary more accurately—splenic images with damaged and undamaged red cells are shown in *Figure 7.1*. The normal spleen contains 5 per cent or less of the red cell volume (Pettit, 1977).

Quantification of ^{51}Cr external counting

As discussed in Chapter 6, one of the major drawbacks of the simple external counting procedure is that the sensitivity of the equipment varies from patient to patient and from organ to organ within each patient. Because of this, if quantitative results are desired, it is necessary to calibrate the detector system for each organ in each patient. All the methods which have so far been developed for performing this calibration involve quantitative imaging. Two methods are currently in use (Williams *et al.*, 1974; Bowring *et al.*, 1975).

In the method described by Williams *et al.* (1974), at the end of the conventional external counting study quantitative images of the spleen are obtained before and after the injection of heat-damaged red cells labelled with 7 MBq of ^{51}Cr. From these images the uptake of the heat-damaged cells in the spleen can be calculated and this information used to calibrate the external counter using the expression in equation (1).

$$P = P' \cdot U \cdot A'/A \ (U' - U) \tag{1}$$

P and P' are the percentage uptakes in the spleen of the initial and calibration doses of ^{51}Cr (P' is obtained from the quantitative imaging), U and U' are the excess counts measured with external counting equipment before and after adminstration of the calibration dose, and A and A' are the amounts of radioactivity administered. The procedure is illustrated in *Figure 7.2*. This method has the disadvantage that it can be applied only to measurements on the spleen.

The principle of the other method is very similar but uses ^{113m}In labelled colloid for the calibration. Since the colloid is taken up in

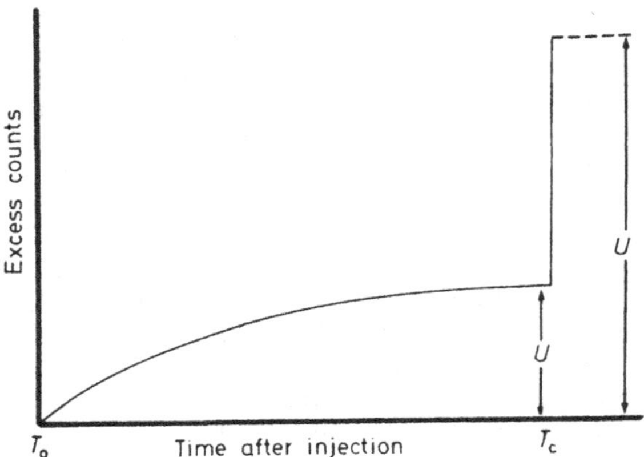

Figure 7.2 The procedure for calibrating ^{51}Cr splenic external counting results using ^{51}Cr-labelled heat-damaged red cells. (At T_0 ^{51}Cr-labelled red cells are injected and conventional external counting is continued until T_c when ^{51}Cr-labelled heat-damaged red cells are injected and the increase in the external countrate is measured)

both the spleen and the liver it may be used to calibrate the external counting data for both organs. ^{113m}In can be used instead of ^{51}Cr because the energies of the γ-rays which the two radionuclides emit are sufficiently similar (*see Table 1.1*). In addition, because of the short half-life of ^{113m}In the calibration can be performed at any time before, during or after the external counting study. In practice it is usually done 24 hours before the start of external counting as it may then also be used to optimize the external counting positions. The uptake of the colloid in each organ is obtained in the standard way and used to calibrate the external detectors using the expression in equation (2).

$$P = P' \cdot U \cdot A' \cdot C / U' \cdot A \tag{2}$$

Where P is the percentage uptake of ^{51}Cr in the organ, P' is the percentage uptake of ^{113m}In in the organ, U is the excess count due to the uptake of ^{51}Cr, U' is the count due to the uptake of ^{113m}In, A and A' are the amounts of ^{51}Cr and ^{113m}In adminstered and C is a constant relating the sensitivity of the external counting equipment to the γ-rays emitted by the two radionuclides. Additionally, if it is possible to carry out whole-body counting on the patient during the external counting study, the use of this calibration method allows an estimate to be made of the uptake of ^{51}Cr in the rest of the reticuloendothelial system by subtracting the amounts in the liver, spleen and circulation from the amount measured to be present in the whole body. However, it must be remembered that the amount of ^{51}Cr present in each organ represents a combination of both the number of red cells destroyed in the organ and the amount of ^{51}Cr subsequently lost from the organ as these cells are broken down. The measured uptake is therefore always less than or equal to the amount of ^{51}Cr in the cells destroyed in the organ.

Two techniques are available to correct for this loss of ^{51}Cr from the organ. If the ^{51}Cr calibration technique is used, further quantitative images can be recorded on the days following the injection of the heat-damaged cells so that the rate of loss can be measured (Williams *et al.*, 1972). If the ^{113m}In calibration technique is used the quantified external counting data can be fitted to a theoretical model and the rates of destruction and loss in each organ can then be obtained (Bowring *et al.*, 1975). Neither of these methods is ideal however, the first because the rate of loss of ^{51}Cr from previously damaged cells may not be equal to the rate of loss from naturally destroyed cells and the second because use of the model may not always be valid. The rate of loss is aproximately 6 per cent/day of the ^{51}Cr in the organ.

Quantification of ferrokinetic external counting

The major interest in external counting with radioisotopes of iron is usually to determine the extent of any extramedullary erythropoiesis that occurs in the spleen. Because of this the quantitative measurement of the amount of iron uptake in organs other than the spleen has not been investigated so fully, although some work has been reported (Steere *et al.*, 1979) and the quantitative imaging technique can in principle be applied equally well to them. The radionuclide of choice for obtaining quantitative images is ^{52}Fe and the procedure is the same as that already described in this chapter. The measurement of the uptake of iron is not, however, a measure of the amount of erythropoiesis that occurs and in order to estimate this it is necessary, because of the short half-life of ^{52}Fe, to perform ^{59}Fe external counting in addition. If it is then assumed that the ^{52}Fe and ^{59}Fe are both metabolized in an identical way the measurement of the uptake of ^{52}Fe at a known time after injection can be used to provide a spot calibration of the ^{59}Fe uptake curve at the same time after injection. The amount of effective erythropoiesis that occurs in the organ can then be estimated from the drop in the ^{59}Fe external count rate over the following days.

In spleen studies, quantitative imaging of ^{52}Fe in patients with myeloproliferative disorders has shown that there is no clear correlation between the splenic uptake of iron and the spleen size, the degree of anaemia, the overall effectiveness of erythropoiesis or the increase in the count rate detected with simple external detectors (Pettit *et al.*, 1976; Pettit, 1977). These studies found that in patients with uncomplicated polycythaemia there was no significant uptake of iron by the spleen, while in myelofibrosis the uptake could be as high as 50 per cent although this is mostly ineffective as far as erythropoiesis is concerned. Because of this Pettit (1977) has suggested that, in patients with myelofibrosis, splenectomy should be considered when quantitative studies show that the haemolytic activity of the spleen predominates over its erythropoietic activity.

References

ARIMIZU, N. and MORRIS, A. C. (1969) Quantitative measurement of radioactivity in internal organs by area scanning. *Journal of Nuclear Medicine*, **10**, 265–269

BATEMAN, S., LEWIS, S. M., NICHOLAS, A. and ZAAFRAN, A. (1978) Splenic red cell pooling: a diagnostic feature in polycythaemia. *British Journal of Haematology*, **40**, 389–396

BOWRING, C. S., FERRANT, A. E., GLASS, H. I., LEWIS, S. M. and SZUR, L. (1975) Quantitative measurement of splenic and hepatic red cell destruction. *British Journal of Haematology*, **31**, 467–477

CHAUDHURI, T. K., ENKHART, J. C., DeGOWIN, R. L. and CHRISTIE, J. H. (1974) ^{59}Fe whole body scanning. *Journal of Nuclear Medicine*, **15**, 667–673

CHRISTENSEN, B. E. (1971) A new method of estimating splenic erythrocyte and plasma volume combined with quantitation of splenic iron incorporation. *Scandinavian Journal of Haematology*, **8**, 245–249

COLEMAN, R. E. and WELCH, D. (1980) Possible pitfalls with clinical imaging of indium-111 leukocytes: concise communication. *Journal of Nuclear Medicine*, **21**, 122–125

DANPURE, H. J., OSMAN, S. and HESSELWOOD, I. P. (1979) Cell damage associated with indium-111 oxine labelling of a human tumour cell line (HeLa 53). *Journal of Labelled Compounds: Radiopharmaceuticals*, **16**, 116

DAVIS, H. H., HEATON, W. A. L., SIEGEL, B. A., MATHIAS, C. J., JOIST, J. H., SHERMAN, L. A. and WELCH, M. J. (1978) Scintigraphic detection of atherosclerotic lesions and venous thrombi in man by In-111 labelled autologous platelets *Lancet*, **i**, 1185–1187

FERRANT, A. and CAUWE, F. (1979) Quantitative organ uptake measurement with a gamma camera. *European Journal of Nuclear Medicine*, **4**, 223–229

FLEMING, J. S. (1979) A technique for the absolute measurement of activity using a gamma camera and computer. *Physics in Medicine and Biology*, **24**, 176–180

GLASS H. I., DeGARRETA, A. C., LEWIS, S. M., GRAMMATICOS, P. and SZUR, L. (1968) Measurement of splenic red blood cell mass with radioactive carbon monoxide. *Lancet*, **i**, 669–670

HEGDE, U. M., WILLIAMS, E. D., LEWIS, S. M., SZUR, L., GLASS, H. I. and PETTIT, J. E. Measurement of splenic red cell volume and visualisation of the spleen with ^{99m}Tc. *Journal of Nuclear Medicine*, **14**, 769–771

KNOSPE, W. H., RAYUDU, V. W. S., CARDELLO, M., FRIEDMAN, A. M., and FORDHAM, E. W. (1976) Bone marrow scanning with 52-iron. Regeneration and extension of marrow after ablative doses of radiotherapy. *Cancer*, **37**, 1432–1442

LAVENDER, J. P., GOLDMAN, J. M., ARNOT, R. N. and THAKUR, M. L. (1977) Kinetics of Indium-111 labelled lymphocytes in normal subjects and patients with Hodgkin's disease. *British Medical Journal*, **2**, 797–799

LEWIS, S. M., CATOVSKY, D., HOWS, J. M. and ARDALAN, B. (1977) Splenic red cell pooling in hairy cell leukaemia. *British Journal of Haematology*, **35**, 351–357

LILIEN, D. J., BERGER, H. G., ANDERSON, D. P. and BENNETT, L. R. (1973) ^{111}In chloride: new agent for bone marrow imaging. *Journal of Nuclear Medicine*, **14**, 184–186

McINTYRE, P. A., LARSON, S. M., EIKMAN, E. A., COLMAN, M., SCHEFFER, V. and HODKINSON, B. A. (1974) Comparison of the metabolism of iron labelled transferrin and indium labelled transferrin by the erythropoietic marrow. *Journal of Nuclear Medicine*, **15**, 856–862

MERRICK, M. V., GORDON-SMITH, E. C., LAVENDER, J. P. and SZUR, L. (1975) A comparison of ^{111}In with ^{52}Fe and ^{99m}Tc sulphur colloid for bone marrow scanning. *Journal of Nuclear Medicine*, **16**, 66–68

PARMENTIER, C., THERAIN, F., CHARBORD, P. AUBERT, B. AND MORARDET, N. (1977) Comparative study of ^{111}In and ^{59}Fe bone marrow scanning. *European Journal of Nuclear Medicine*, **2**, 89–92

PERRY. M. C. and HOLMES, R. A. (1980) The radioindium bone-marrow image in acute (malignant) myelofibrosis. *Journal of Nuclear Medicine*, **21**, 139–141

PETERS, A. M. and LAVENDER, J. P. (1980) White cell scanning in some difficult diagnostic situations. *Nuclear Medicine communications*, **1**, 180

PETTIT, J. E. (1977) Spleen Function. *Clinics in Haematology*, **6**, 639–656

PETTIT, J. E., LEWIS, S. M., WILLIAMS, E. D., GRAFTON, C. A., BOWRING, C. S. and GLASS, H. I. (1976) Quantitative studies of splenic erythropoiesis in polycythaemia and myelofibrosis. *British Journal of Haematology*, **34**, 465–475

PETTIT, J. E., LEWIS, S. M. and NICHOLAS, A. W. (1979) Transitional myeloproliferative disorder. *British Journal of Haematology*, **43**, 167–184

PETTIT, J. E., WILLIAMS, E. D., GLASS, H. I., LEWIS, S. M., SZUR, L. and WICKS, C. J. (1971) Studies of splenic function in the myeloproliferative disorders and generalised malignant lymphomas. *British Journal of Haematology*, **20**, 575–586

PRYOR, D. S. (1967) The mechanisms of anaemia in tropical splenomegaly. *Quarterly Journal of Medicine*, **36**, 337–356

RAMCHANDRAN, T., MARGOULEFF, D. and ATKINS, H. (1980) Spleen scanning in humans with ^{99m}Tc labelled erythrocytes. *Journal of Nuclear Medicine*, **21**, 13–16

RAYUDU. G. V. S., SHIRAZI, S. P. H., FORDHAM, E. W. and FRIEDMAN, A. M. (1973) Comparison of the use of ^{52}Fe and ^{111}In for haemaopoietic marrow scanning. *International Journal of Applied Radiation and Isotopes*. **24**, 451–454

RONAI, P., WINCHELL, H. S., ANGER, H. O. and LAWRENCE, J. H. (1969) Whole body scanning of ^{59}Fe for evaluating body distribution of erythropoietic marrow, splenic sequestration of red cells and hepatic deposition of iron. *Journal of Nuclear Medicine*, **10**, 469–474

SEGAL, A. W., DETEIX, P., GARCIA, R., TOOTH, P., ZANELLI, G. D. and
ALLISON, A. C. (1978) Indium-111 labelling of leukocytes: A detrimental
effect on neutrophil and lymphocyte function and an improved method of cell
labelling. *Journal of nuclear Medicine*, **19**, 1238–1244

SINN, H. and SILVESTER, D. J. (1979) Simplified cell labelling with Indium-111
acetylacetone. *British Journal of Radiology*, **52**, 758–759

STEERE, H. A., LILLICRAP, S. C., CLINK, H. M. and PECKHAM, M. J.
(1979) The recovery of iron uptake in erythropoietic bone marrow following
large field radiotherapy. *British Journal of Radiology*, **52**, 61–66

THAKUR, M. L., NUNN, A. D. and WATERS, S. L. (1971) Iron 52, improving
its recovery from cyclotron targets. *International Journal of Applied Radiation
and Isotopes*, **22**, 481–483

TOGHILL, P. J. (1964) Red cell pooling in enlarged spleens. *British Journal of
Haematology*, **10**, 347–357

VAN DYKE, D., SHKURKIN, C., PRICE, D., YANO, Y. and ANGER, H. O.
(1967) Difference in distribution of erythropoietic and reticuloendothelial marrow in haematological disease. *Blood*, **30**, 364

WILLIAMS, E. D., SZUR, L., GLASS, H. I., LEWIS, S. M., PETTIT, J. E. and
AHUJA, S. (1974) Measurement of red cell destruction in the spleen. *Journal of
Laboratory and Clinical Medicine*, **84**, 134–146

WILLIAMS, E. D., AHUJA, S., SZUR, L., LEWIS, S. M. and GLASS, H. I.
(1972) Rate of loss of ^{51}Cr from the spleen. *Journal of Nuclear Medicine*, **13**,
686–688

8

Absorption, loss and clearance studies

The use of radionuclide tracers for absorption measurements is an obvious and simple application of their properties and in haematology these measurements have found universal acceptance for the measurement of vitamin B_{12} absorption and iron absorption. There are several ways in which absorption can be measured and, since some of these involve measurements of the loss of tracer from the body or the clearance of tracer from the blood, the measurement of blood loss and the clearance rate of heat-damaged red cells from the circulation are also discussed in this chapter.

Vitamin B_{12} absorption

In patients suffering from vitamin B_{12} deficiency it is necessary to determine whether the subject is able to absorb the vitamin normally. There are only two basic reasons for malabsorption and it is necessary to be able to distinguish between these. A conventional mixed diet supplies approximately $5\,\mu g$ of the vitamin each day. In the stomach the vitamin is bound to intrinsic factor which is normally present in vastly greater amounts than is actually required (generally only 1–2 per cent of the available intrinsic factor is used). The complex formed by the vitamin B_{12} with the intrinsic factor then becomes attached to receptor sites in the ileum and the vitamin alone enters the portal blood attached to a specific carrier protein (transcobalamin). Finally the B_{12} is taken up in the liver where it has a slow turnover with a biological half-life of about one year. The maximum amount of B_{12} that can be absorbed from a single meal is limited to $1.5–2.0\,\mu g$ and because of this test doses are normally $0.25–1.0\,\mu g$. When doses of this size are administered to normal subjects about 50 per cent is absorbed and if larger test doses are used the percentage absorbed falls. The absorbed vitamin appears in the blood 2–3 hours after oral administration and accumulation in the liver starts immediately. Peak blood levels are found at 8–12 hours and the maximum level in the liver is usually reached by 24 hours. Malabsorption is generally due either to lack of intrinsic factor or to impaired intestinal absorption of the B_{12}–intrinsic factor complex. In order to distinguish between these two states two measurements are usually made. First, an absorption measurement is made with a dose of labelled vitamin B_{12} alone, and secondly, if this is not absorbed adequately, the test is repeated with a dose of labelled vitamin B_{12} complexed with intrinsic factor.

Five different radionuclide tracer techniques have been used for these measurements and they are all briefly discussed below, although only two of them, the whole body counting method and the urinary excretion method, are now widely used. In all these tests $1.0\,\mu g$ of labelled vitamin B_{12} is administered orally to the patient who should have been fasted overnight before receiving the dose and then should take no food for a further 2 hours afterwards . A standard must be prepared for most measurements; it is generally made up at the same time as the test dose so that the relative amounts of the tracer in the standard and test dose may be simply determined. Four radioisotopes of cobalt have been used for labelling the vitamin B_{12} — ^{56}Co, ^{57}Co, ^{58}Co and ^{60}Co—but of these only ^{57}Co and ^{58}Co are now commonly used because the radiation dose to the pateient from use of the others is much larger (*see Table 1.2* for a summary of their properties). The labelled material, both in a pure form and in a complex with intrinsic factor, is obtained from commercial suppliers and is normally supplied so that $1\,\mu g$ of B_{12} contains 25–50 kBq of radioactive tracer.

Faecal excretion method

This method involves the collection of all faeces passed after the administration of the test dose until the amount of radioactivity collected in a 24-hour period is less than 2 per cent of the dose administered. This generally takes at least 4–6 days. Each faecal sample is counted in a large-volume sample counter (*see* Chapter 2) and compared with a dilution of the standard placed in a similar container, in order that the total amount of labelled B_{12} collected in each sample may be simply determined. The amount absorbed is then given by the difference between the amount administered and the total amount collected in the faeces.

The normal absorption is $0.5\,\mu g$ or more but patients with pernicious anaemia (lack of intrinsic factor) or intestinal malabsorption usually absorb less than $0.2\,\mu g$ (Mollin and Waters, 1971). By repeating the test with intrinsic factor added to the test dose the two causes of low absorption can be distinguished: the absorption becomes normal in pernicious anaemia but remains low if there is an intestinal absorption defect.

Although this is an indirect method for measuring absorption, the technique was the first used and is capable of reasonable accuracy provided great care is taken to ensure that the stool collection is complete. This is very difficult to ensure unless the patient is in a metabolic unit. If the collection is incomplete the absorption will be overestimated—an error which could be dangerous. Because of this, and also because of the general unpleasantness for both patient and technician of collecting and handling faecal samples, it is not commonly used.

External counting method

This method relies on the fact that most of the administered B$_{12}$ normally accumulates in the liver. An external detector is positioned over the liver in the same way as described in Chapter 6 for ^{51}Cr external counting and measurements are made until a steady level is maintained—this generally takes 3–7 days. The method suffers from all the same errors as simple ^{51}Cr external counting although the detector can be simply calibrated by injecting the patient intravenously with labelled B$_{12}$ and comparing the increase in count rate which this produces with the original count rate before injection. The detected count rates are always low, which makes the method rather time consuming. In addition, low results are of doubtful significance in severe liver disease since the disease can interfere with the ability of the liver to take up the absorbed B$_{12}$. Because of this the method has now been virtually abandoned.

Whole body counting method

If the patient can be counted in a whole body counter (*see* Chapter 2), it is possible to obtain an accurate measurement of the quantity absorbed provided that all the unabsorbed material has been excreted. Where whole body counters are available this method is undoubtedly the best for measuring absorption, and its only disadvantage is the necessity to wait for up to one week or occasionally more for the result. This time delay is necessary in order to allow all the unabsorbed radioactivity to be excreted. A background count is measured with the patient in the counter shortly before the labelled vitamin B$_{12}$ is administered and a second count is made soon after; the difference corresponds to 100 per cent of the dose. A third measurement is done one week later with the patient in an identical position and the relationship between the patient background and the machine background (previously determined) is assumed to be the same so that the amount absorbed may be simply calculated. It is also necessary to count a standard of the cobalt radionuclide on each occasion in order that corrections can be made for any changes in the detector sensitivity and for decay of the radionuclide. It is not necessary for the relative quantities of the radionuclide in the administered dose and standard to be known as the initial measurement on the patient gives the 100 per cent value. Vitamin B_{12} labelled with ^{58}Co is normally used for whole-body-counter measurements since the response of the instrument is less sensitive to changes in distribution of the tracer within the body with this radionuclide than it is with ^{57}Co. This is due to the higher energy of the γ-rays emitted by ^{58}Co. However the successful use of ^{57}Co has also been recently described (Tait and Hesp, 1976; Smith and Hesp, 1979).

Plasma radioactivity method

As has already been noted, the level of radioactivity in plasma generally reaches a peak 8–12 hours after oral adminstration of the test dose; at this time the normal plasma radioactivity is 0.7–2.2 per cent of the administered dose per litre of plasma (Coupland, 1966). The method simply involves taking a blood sample 8 hours after administration of the labelled B_{12} and determining the percentage present in one litre of plasma. It is adequate as a rapid test to determine whether or not absorption is normal, but it does not give an accurate measurement of absorption. Discrepancies have also been reported between this test and the Schilling Test (*see* below) in patients with malabsorption (McIntyre and Wagner, 1966).

Urinary excretion method (Schilling test)

This method (Schilling, 1953) relies on the administration of a large 'flushing' dose of non-radioactive vitamin B_{12} given intramuscularly to saturate the transcobalamins so that the radioactive vitamin B_{12}, which has been given orally, and which has been absorbed, may be largely excreted in the urine instead of being transported to the liver. This 'flushing' dose is usually 1000 μg and must be given within 2 hours of the administration of the oral dose. The urine excreted over the next 24 hours is then collected and the total amount of the radioactive B_{12} excreted in the urine is measured. If a large-volume sample counter is available the entire urine specimen can be counted but if not an aliquot can be counted in a normal sample counter and the amount in this multiplied by the ratio of the total urine volume collected to the aliquot volume. Normal excretion is greater than 7–8 per cent of the orally administered dose and is less than 3 per cent in patients with pernicious anaemia or intestinal malabsorption (Mollin and Waters, 1971), although results between these figures are quite often found. The test can be repeated with the addition of intrinsic factor 48 hours after the start of the first test provided an additional 'flushing' dose of non-radioactive B_{12} is given 24 hours before.

The method provides a relatively quick result and is the most widely used of the available tests for B_{12} absorption. The results from it are normally reliable as long as the urine collection is complete (Chanarin and Waters, 1974) and the patient does not suffer from any renal disease, which may cause falsely low results to be obtained. The need for the large 'flushing' doses is also a disadvantage as these may interfere with other studies, and failure to administer the 'flushing' dose will lead to falsely low results. In patients with known renal disease urine collection may be continued for 48 hours with a second flushing dose administered at 24 hours (in normal subjects this second collection usually contains approximately one-third of the radioactivity present in the first sample). Alternatively, and more accurately, if the patient's creatinine clearance is measured a correction to the measured percentage excreted can be made by multiplying the measured percentage by the ratio of the normal creatinine clearance to the patient's measured creatinine clearance.

A time saving modification of the basic method (Katz, DiMase and Donaldson, 1963) combines both stages of the test by administering the unbound B_{12} and the intrinsic-factor bound B_{12} simultaneously. The unbound B_{12} is usually labelled with ^{58}Co and the bound with ^{57}Co. Since the γ-rays emitted by these two radionuclides are easily distinguished (*see Table 1.2*), it is possible to determine the amounts of each radionuclide in the single urine collection by appropriate setting of the pulse height analyser windows of the sample counter (*see* Chapter 2). In normal subjects equal and normal amounts of both radionclides are excreted, in patients with intestinal malabsorption equal but reduced amounts of the two radionuclides are excreted and in patients with pernicious anaemia reduced amounts of the unbound tracer and normal amounts of the bound tracer are excreted. This modification of the test also has the advantage therefore that by looking at the ratio of the amounts of bound to unbound tracer that are excreted it is possible to distinguish the cases of pernicious anaemia even when the urine collection is incomplete. The normal range of the ratio of the amounts of bound to unbound tracer excreted is 0.8–1.3 and in pernicious anaemia the ratio is greater than 2.0. A typical calculation of the results is shown in *Table 8.1*. This modified test, in which both the unbound and bound tracers

Table 8.1 Calculation of the percentages of bound and unbound vitamin B_{12} excreted in the urine when these are administered simultaneously

Volume of urine collected in 24 hours	= 1370 ml
Volume of sample counted	= 20 ml
Ratio of standard sample to quantities administered	= 1:50

Sample count rates correct for background(counts/s)

	^{58}Co (unbound)	^{57}Co (bound)	^{57}Co (corrected for cross-talk)
^{58}Co standard	93.57	7.80	–
^{57}Co standard	–	208.81	208.81
Urine	14.62	34.90	33.68*

*Cross-talk correction = $14.62 \times 7.80/93.57 = 1.22$
Percentage of ^{58}Co excreted = $(1370 \times 14.62 \times 100)/(93.57 \times 50 \times 20) = 21.4$
Percentage of ^{57}Co excreted = $(1370 \times 33.68 \times 100)/(208.81 \times 50 \times 20) = 22.1$
Ratio of bound:unbound excreted = $22.1:21.4 = 1.03$

are administered simultaneously, is the basis of the commonly used and commercially available 'Dicopac' test kit (available from The Radiochemical Centre, Amersham, England). It should be mentioned, however, that if the two tracers supplied with this test kit are not administered simultaneously misleading results may be obtained.

Iron absorption

Iron absorption has been studied in great detail and is of major importance in understanding the disorders of iron metabolism. However, in individual patients the measurement has only proved to be of limited value because iron absorption varies greatly with the chemical form and concentration in which it is administered. In addition, since the consumption of different food also affects iron absorption, absorption of inorganic iron by the patient will not necessarily mean that the patient can absorb iron from his/her diet. The whole subject has been recently reviewed by Cook and Lipschitz (1977). As a result of these complications the most common clinical use of the measurement of iron absorption is to demonstrate normal absorption in patients suspected of having an absorption defect.

Three methods have been used to make the measurement: a direct measurement by whole body counting; a measurement of the amount excreted in the faeces, and a dual radionuclide technique which involves measuring the relative quantities of orally and intravenously administered iron that are incorporated in the circulating red cells. In the past tests have been made with the radioactive iron in numerous chemical forms but the most commonly used of these is now the ferrous form (Dacie and Lewis, 1975). Approximately $50\,kBq$ of ^{59}Fe in the form of ferric chloride (specific activity $\approx 50\,kBq/\mu g$) is added to $5\,mg$ of carrier iron in the form of aqueous ferrous sulphate and $50\,mg$ of ascorbic acid. The last-mentioned reduces the ferric chloride to the ferrous state. This solution is made up to a volume of about $20\,ml$ with water and a known fraction is retained for use as a standard. The remainder is administered orally to the patient who should have previously been fasted over night and should then continue to fast for a further two hours. Normal subjects usually absorb 10–35 per cent of such a test dose although when iron deficient the absorption can be as high as 80 per cent (Dacie and Lewis, 1975).

Faecal counting method

This method is identical to that described for the measurement of vitamin B_{12} absorption and suffers from the same disadvantages. In addition it is necessary to collect the stool samples for longer than in the B_{12} test; 7–10 days are usually required.

Whole body counting method

This is also basically the same as that described for the measurement of B_{12} absorption except that the final uptake measurement needs to be done 10 days after administration of the iron. A problem that sometimes arises with whole-body iron measurements is that the sensitivity of the counter may vary with the distribution of the radionuclide in the body. Since the distribution varies from being concentrated in the stomach at the time of the intial measurement to being spread throughout the whole body later, this is potentially a greater problem with iron than it is with B_{12} in which the change in distribution from stomach to liver is much less. The magnitude of the error that this can introduce depends on the design of the whole body counter that is used but is generally smallest for the 'shadow shield' type (*see* Chapter 2).

Dual tracer method

In most patients a high proportion of the radioiron absorbed from an oral test dose may be expected to be incorporated in the forming red cells and then to appear in the circulation. Therefore, by measuring the amount of the test dose incorporated in the red cells 10–14 days after administration, it is possible to determine the amount absorbed. In order to do this accurately two tracers are used, ^{55}Fe is used to label the orally adminstered test dose and ^{59}Fe-labelled transferrin is injected intravenously as described in Chapter 5. A blood sample is taken 10–14 days later and the relative amounts of the two tracers in the red cells is determined. The amount of ^{59}Fe is taken as representing 100 per cent absorption and hence the absorption of the ^{55}Fe test drink can be determined using the expression in equation (1).

$$\% \text{ absorption of oral dose} = \frac{(\% \text{ of oral dose/ml red cells}) \times 100}{\% \text{ of intravenous dose/ml red cells}} \tag{1}$$

Unfortunately, this procedure has three disadvantages. First, it is assumed that the distribution of orally administered iron in the body will be the same as that of the intravenously administered iron and this may not be the case in patients with transferrin saturation. Secondly, the additional radiation dose to the patient with the ^{55}Fe may not be felt to be justified, and thirdly, sample preparation is more complex than for counting ^{59}Fe alone. This last problem is due to the very low energy of the radiation emitted in the decay of ^{55}Fe and, because of this, liquid scintillation counting with its associated difficulties (*see* Chapter 2) has to be used for this radionuclide. Because of these factors this method is not as often used as the previous two methods described and of these the whole body counting method is normally preferred if the equipment is available.

Measurement of blood loss

The quantitative measurement of blood loss due to haemorrhage into the gastrointestinal tract is occasionally of use and can be simply assessed by determining the amount excreted in the faeces using ^{51}Cr labelled red cells. The patient's own red cells are labelled in the standard way (*see* Appendix 1) with 3–4 MBq of ^{51}Cr and are injected intravenously. It is not necessary to prepare a standard. All stools are collected for 5 days so that the daily blood loss may be calculated. A blood sample is taken on each day of the study and a known volume of it is diluted with water into the same type of container as is used for the stool collection. The stool samples are then counted in a large-volume sample counter, the count rate obtained is compared with that of the corresponding blood sample and thus the volume of blood in the sample is simply obtained. Using this method blood losses of as little as one ml/day are easily detectable. It is essential that great care is taken to ensure that the stools are not contaminated with urine since the ^{51}Cr, which is eluted from the red cells (*see* Chapter 4), is excreted in the urine and therefore urine contamination can lead to an overestimate of the volume of blood in the sample. This is usually a problem only if the volume of urine contaminating a 24-hour collection is greater than about 10ml.

Recently the use of red blood cells labelled with ^{99m}Tc has also been described for the detection of gastrointestinal bleeding although not for the measurement of the actual volume of blood lost (Winzelberg *et al.*, 1979). The patient's red cells were labelled with 750MBq of ^{99m}Tc and images of the abdomen were obtained with a gamma camera at 5 minute intervals for 30 minutes after injection and at 1, 2 and 24 hours. When focal accumulations of activity were noted in regions of the abdomen normally free of radioactivity, this was taken as evidence of bleeding. False positive results were obtained in some patients in whom there was initial gastric accumulation of ^{99m}Tc and it was suggested that this source of error could be minimized by keeping the patients on continuous nasogastric suction during the 24-hour imaging period. It is not yet known how sensitive this technique is but it is certainly not capable of the same sensitivity as the ^{51}Cr method.

One report has also been made of the use of ^{111}In-labelled red cells for the detection of sites of gastrointestinal bleeding (Ferrant *et al.*, 1980) and, provided this technique is validated, it would appear to be a superior imaging method to that using [99m]Tc.

Clearance of heat-damaged red cells from the blood

The measurement of the clearance rate of deliberately damaged red cells from the circulation has become an established method for obtaining information on splenic function. In the past ^{51}Cr-labelled heat-damaged cells or [197]Hg-labelled MHP-damaged cells have been used for this test but it is now commonly performed only with ^{99m}Tc-labelled heat-damaged cells, blood samples being taken following injection of the labelled damaged cells. It is important that the damaging is done in a carefully controlled and standardized way if consistent results are to be obtained and it is also important that the timing of the first blood sample after injection of the cells is not delayed. The cells are damaged and labelled with ^{99m}Tc in the way described in Appendix 1. The first blood sample should be taken exactly 3 minutes after the midpoint of the injection with further samples being taken at 10, 20, 30 and 60 minutes. The quantity of radioactivity per millilitre of whole blood in each sample is then measured and expressed as a percentage of the quantity in the first sample. The results are plotted on semilogarithmic graph paper and the half-clearance time obtained from the graph as shown in *Figure 8.1*. The normal half-clearance time for heat-damaged cells is 5–15 minutes (Pettit, 1977) but is slower if MHP is used.

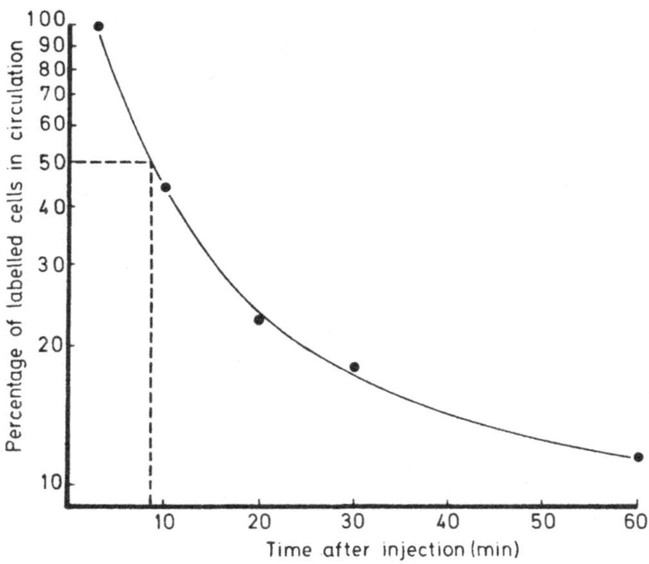

Figure 8.1 The clearance of heat-damaged red cells from the circulation in a normal subject. The half-clearance time is approximately 9 minutes

It has been shown by Fischer *et al.*, (1971) that the clearance curve has two components, a faster component which is due to clearance into the spleen and a slower component due to clearance into the rest of the reticuloendothelial system. It is because of this that the reproducibility of the damaging technique and the timing of the first blood samples are so important. If the first sample is more significant and the measured half-clearance time will be longer than it should be. Similarly, if the damaging is not performed in a reproducible way, the relative amounts of the cells cleared by the spleen and by the rest of the reticuloendothelial system will alter and affect the measured clearance time.

Because of these two sources of error in the simple measurement it has been suggested by Bowring, Glass and Lewis (1976) that only the splenic component of the clearance should be measured. This can be done either by sequential imaging of the spleen (ideally with a gamma camera linked to a computer) and calculation of the half-clearance time into the spleen of the damaged cells, or by splitting the blood clearance curve into its two components and taking the half-clearance time of the faster component (Fischer *et al.*, 1971). The normal half-clearance time using these methods is less than 7 minutes.

These methods are undoubtedly capable of greater accuracy than the simple method but require considerably more effort on the part of the technicians doing the measurements. The simple method is probably adequate for routine use provided that the necessary care is taken with the damaging of the cells and the timing of the samples.

References

BOWRING, C. S., GLASS, H. I. and LEWIS, S. M. (1976) Rate of clearance by the spleen of heat-damaged erythrocytes. *Journal of Clinical Pathology*, **29**, 852–854

CHANARIN, I. and WATERS, D. A. W. (1974) Failed Schilling Tests. *Scandinavian Journal of Haematology*, **12**, 245–248

COOK. J. D. and LIPSCHITZ, D. A. (1977) Clinical measurements of iron absorption. *Clinics in Haematology*, **6**, 567–581

COUPLAND, W. W. (1966) Plasma radioactivity after radioactive vitamin B_{12} given orally. *Medical Journal of Australia*, **1**, 1020–1023

DACIE, J. V. and LEWIS, S. M. (1975) *Practical Haematology*, 5th Edn. pp. 446–447. Edinburgh and London: Churchill Livingstone

FERRANT, A., DEHASQUE, N., LENERS, N. and MEUNIER, H. (1980) Scintigraphy with In-111 labelled red cells in intermittent gastrointestinal bleeding. *Journal of Nuclear Medicine*, **21**, 844–845

FISCHER, J., WOLF, R., MUNDSCHENK, H., LEON, A. and HROMEC, A. (1971) Clinical value of the function investigation of the spleen using ^{51}Cr labelled heat damaged erythrocytes. In *Dynamic Studies with Radioisotopes*. pp. 445–462. Vienna: International Atomic Energy Agency

KATZ, J. H., Di MASE, J. and DONALDSON, R. M. (1963) Simultaneous administration of gastric juice bound and free radioactive cyanocobalamin: Rapid procedure for differentiating between intrinsic factor deficiency and other causes of vitamin B_{12} malabsorption. *Journal of Laboratory and Clinical Medicine*, **61**, 266–271

McINTYRE, P. A. and WAGNER, H. N. (1966) Comparison of the urinary excretion and 8 hour plasma tests for vitamin B_{12} absorption. *Journal of Laboratory and Clinical Medicine*, **68**, 966–971
MOLLIN, D. L. and WATERS, A. H. (1971) Absorption and metabolism of vitamin B_{12} and folic acid. In *Radioisotopes in Medical Diagnosis*. pp. 412–436 Eds. Belcher, E. H. and Vetter, H. London: Butterworths
PETTIT, J. E. (1977) Spleen Function. *Clinics in Haematology*, **6**, 639–656
SCHILLING, R. F. (1953) Intrinsic Factor Studies II. The effect of gastric juice on the urinary excretion of radioactivity after the oral administration of radioactive B_{12}. *Journal of Laboratory and Clinical Medicine*, **42**, 860
SMITH, T. and HESP, R. (1979) Measurement of ^{57}Co labelled vitamin B_{12} using a liquid scintillator whole body counter. *British Journal of Radiology*, **52**, 832–835
TAIT, C. E. and HESP, R. (1976) Measurement of ^{57}Co vitamin B_{12} uptake using a static whole body counter. *British Journal of Radiology*, **49**, 948–950
WINZELBERG, G. G., McKUSICK, K. A., STRAUSS, H. W., WALTMAN, A. C. and GREENFIELD, A. J. (1979) Evaluation of gastrointestinal bleeding by red cells labelled in vivo with technetium-99m. *Journal of Nuclear Medicine*, **20**, 1080–1086

Labelling procedures

In this appendix some of the commonly used labelling procedures are described. Where the ICSH has specifically approved a method of labelling this is described but in the cases where no standardized method has been agreed the methods described are ones which have been used and found reliable in several independent establishments but are not necessarily the most widely used. The following procedures are described:

^{51}Cr labelling of red cells
^{51}Cr labelling of heat-damaged red cells
^{51}Cr labelling of platelets
^{99m}Tc labelling of red cells
^{99m}Tc labelling of heat-damaged red cells
^{111}In labelling of platelets
Indium labelling of transferrin
Iron labelling of transferrin

The other radiopharmaceuticals that are mentioned in the main text and which are still in common use are all obtainable commercially, either already labelled or with simple kits to facilitate labelling. It is now possible to obtain ^{111}In-oxine specifically formulated for labelling platelets and white cells.

^{51}Cr labelling of red cells

These two methods have both been recommended by the ICSH (1971) and the relevant elution correction factors for use in red cell survival studies are given in *Table A1.1*. The two methods are the ACD method using acid–citrate–dextrose and the CPD method using citrate–phosphate–dextrose, and of these the ACD method is now accepted as the reference technique (ICSH, 1980).

ACD method

The ACD solution, which can be made in advance and ampouled in convenient volumes, consists of 2.2 g trisodium citrate dihy-

Table Al.1 Elution correction factors for the ACD and CPD ^{51}Cr red-cell labelling methods

Day	Correction factor		Day	Correction factor	
	ACD	CPD		ACD	CPD
0	1.00	1.00	16	1.23	1.22
1	1.01	1.03	17	1.24	1.23
2	1.03	1.05	18	1.26	1.25
3	1.04	1.06	19	1.27	1.26
4	1.06	1.07	20	1.29	1.27
5	1.07	1.08	21	1.31	1.29
6	1.08	1.10	22	1.32	1.31
7	1.10	1.11	23	1.34	1.32
8	1.11	1.12	24	1.35	1.34
9	1.13	1.13	25	1.37	1.36
10	1.14	1.14	26	1.38	1.38
11	1.16	1.16	27	1.40	1.40
12	1.17	1.17	28	1.41	1.42
13	1.19	1.18	29	1.43	1.45
14	1.20	1.19	30	1.44	1.47
15	1.21	1.20			

drate, 8g citric acid, 25g dextrose and water for injection to 1ℓ. The ACD solution is used in the ratio of 1.5 volumes to 10 volumes of whole blood. Assuming that the red cells in 10 ml of blood are to be labelled, 1.5 ml of ACD solution is put in a sterile 20 ml container and the 10 ml of whole blood from the patient is added to it immediately after venipuncture, and the whole gently mixed. It is then centrifuged at 1000–1500 g for 5–10 minutes after which the supernatant plasma is removed. The required quantity of ^{51}Cr sodium chromate is then added—7.5 kBq/kg body weight for blood volume measurement, 25 kBq/kg body weight for red cell survival measurements and 50 kBq/kg body weight for red cell survival with external counting measurements. The volume added should be approximately 0.2 ml—the ^{51}Cr sodium chromate can be diluted with saline if necessary. The mixture is then allowed to stand for 15 minutes either at room temperature or in a water bath at 37 °C. The labelled cells are then washed twice in saline in order to remove any unbound ^{51}Cr and finally resuspended in saline ready for injection.

CPD method

The CPD solution is made up from 30 g trisodium citrate dihydrate, 0.15 g sodium dihydrogen phosphate, 2 g dextrose and water for injection to 1ℓ. The solution has a pH of approximately 6.9 at room temperature and a brown tinge due to caramelization after autoclaving. At least 2 volumes of CPD are used for 1 volume of whole blood. Thus, if 10 ml of blood is taken from the patient it is added to at least 20 ml of CPD immediately after venipuncture. The suspension is then centrifuged and the procedure becomes identical to the ACD method.

^{51}Cr labelling of heat-damaged red cells

The procedure for obtaining ^{51}Cr-labelled heat-damaged red cells is to label the cells as described above and then to damage them. After the cells have been labelled and washed they must be placed in a sterile *glass* bottle (if not already in one) and then placed in a water bath for exactly 20 minutes, with occasional gentle mixing. The temperature of the water bath should not vary outside the range 49.5–50.0°C (Dacie and Lewis, 1975). After heating the cells should be washed again and then resuspended in saline ready for injection. The cells should be labelled with about 25 kBq/kg body weight for clearance studies and 75 kBq/kg body weight for spleen imaging.

^{51}Cr labelling of platelets

This method is that recommended by the ICSH (1977). It is, however, rather more involved than most labelling methods and a centrifuge large enough to allow bags containing 500 ml of blood to be centrifuged at $1500g$ is required. It is essential that this centrifuge has a smooth braking action. Facilities for ensuring that the temperature does not vary beyond the range of 20–25°C are also needed.

The patient is weighed and a platelet count performed and then, from consideration of this information and of the patient's age and clinical condition, an appropriate volume of blood for labelling is decided on (usually 200–500 ml). For every 100 ml of blood that is to be taken, 15 ml of ACD is put into a sterile plastic transfusion bag which must also have two dry satellite bags attached to it. The blood is then taken into the bag and the entire bag is centrifuged at $300g$ for 15 minutes including acceleration time (but not deceleration). The supernatant platelet rich plasma (PRP) is transferred into the first satellite bag and approximately 5 ml of ACD added per 100 ml of PRP bringing the pH of the PRP to 6.5 ± 0.2 (if desired the ACD can be put in the bag first and the PRP added to it). The platelets are then sedimented into a pellet by centrifugation at $1500g$ for 15 minutes. All but 5 ml of the supernatant platelet poor plasma (PPP) is now transferred to the second satellite bag without disturbing the pellet. The platelets in the pellet are resuspended in the remaining 5 ml of PPP by repeated gentle inversion of the bag and the ^{51}Cr sodium chromate added (approximately 50 kBq/kg body weight). The mixture is incubated without shaking for 30 minutes at 20–25 °C. All but 40 ml of the PPP is returned from the second satellite bag to the mixture and 200 ml of filtered air introduced into the bag. The bag is then centrifuged as before in order to re-form the platelet pellet, the PPP is removed, and a further 20 ml of PPP from the second satellite bag is carefully layered over the pellet and then decanted off and discarded. (The object of these stages is to remove any unbound ^{51}Cr.) The labelled platelets are then gently resuspended in the last 20 ml of PPP from the second satellite bag and are ready for injection.

^{99m}Tc labelling of red cells

Many different methods of labelling red cells have been described and these can be broadly separated into two methods, a pre-tinning method and a post-tinning method. The difference between these depends on whether stannous chloride is added to the red cells before or after the ^{99m}Tc. The pre-tinning methods seem in general to be more satisfactory because the uptake of ^{99m}Tc by the cells is usually greater than with the post-tinning methods and the rate of elution of ^{99m}Tc from the red cells after injection is usually less. An example of each method is described. The post-tinning method is that used by Ferrant, Lewis and Szur (1974). The pre-tinning method is that described by Jones and Mollison (1978) and recently approved by the ICSH (1980). With the post-tinning method the loss of ^{99m}Tc from the circulation is negligible for the first 20 minutes after injection but reaches a value of 7 per cent at 1 hour, while with the pre-tinning method this loss is reduced to 4 per cent at 1 hour so that it is seldom necessary to make a correction for it.

Post-tinning method

10 ml of whole blood is collected into 1.5 ml of ACD solution and centrifuged at $1500g$ for 5 minutes. The plasma is then removed and the ^{99m}Tc, as pertechnetate from the generator, is added in a

volume of approximately 0.2 ml of saline. The quantity of ^{99m}Tc added should be about 5 MBq if the cells are being labelled for a blood volume determination and about 200 MBq if an image of the splenic blood pool is to be obtained. The mixture is incubated at room temperature for 15 minutes. A volume of freshly prepared stannous chloride solution (approximately 1 µg of tin/ml of red cells) is then added and the mixture left to stand for a further 5 minutes. The cells are then washed three times with normal saline in order to remove the unbound ^{99m}Tc and resuspended in saline for injection. Details of the stannous chloride solution are given below.

Pre-tinning method

A 5 ml sample of heparinized blood from the patient is added to 20 ml of saline and centrifuged. The supernatant is then removed and a volume of freshly prepared stannous chloride solution is added (approximately 0.015 µg of tin/ml of red cells). The mixture is then left to stand for 5 minutes at room temperature. The ^{99m}Tc is then added in 0.2 ml of saline and left for a further 5 minutes incubation. The quantity of ^{99m}Tc can be reduced to about 70 per cent of that used in the post-tinning method because of the greater labelling efficiency. After incubation the cells are washed in saline and then resuspended ready for injection. Ideally the saline used for washing and resuspension should be ice-cold and the suspension kept on ice until injected.

Preparation of the stannous chloride solution

It is essential that the length of time between dissolving the $SnCl_2$ and adding the solution to the red cells should be as short as possible and it should never exceed 10 minutes. A suitable solution can be made by dissolving 2 mg of $SnCl_2.2H_2O$ in 20 ml of saline and passing this through a 0.22 µm millipore filter. This solution can be used directly in the post-tinning method since it has a concentration of tin of approximately 5 µg/0.1 ml of solution but should be further diluted for the pre-tinning method. If 1 ml of the filtered solution is added to a 0.5 ℓ container of saline a solution of approximately 0.01 µg Sn/0.1 ml of solution will be obtained that is suitable for the pre-tinning method.

^{99m}Tc labelling of heat-damaged red cells

The labelling procedure is identical to that described above for undamaged red cells except that after the supernatant plasma has been removed and before either ^{99m}Tc or stannous chloride is added, the cells are damaged. The damaging is done in the same way as described for ^{51}Cr. After damaging the cells are labelled as described above. For simple blood clearance measurements the cells should be labelled with about 5 MBq of ^{99m}Tc and for spleen imaging with about 50 MBq.

^{111}In labelling of platelets

The method described is that of Heaton *et al.* (1979) and is based on the fact that when ^{111}In is chelated with 8-hydroxyquinoline (oxine) a lipid-soluble complex is formed which labels platelets with high efficiency.

The ^{111}In-oxine complex is prepared immediately before the labelling commences by the addition of 50 µl of oxine in absolute ethanol (1 mg/ml) to the volume of 111-indium chloride in dilute HCl containing the required activity—3 MBq for survival studies and 50 MBq for imaging studies—0.2–0.7 ml. This solution is then mixed and added to 4 ml of modified ACD-saline solution (citric acid 8.0 g/ℓ, sodium citrate 22.4 g/ℓ, anhydrous dextrose 2.0 g/ℓ, diluted with isotonic saline 1:7) following which the pH is adjusted to 6.5 with molar NaOH solution.

A 43 ml sample of blood is drawn from the patient into 7 ml of modified ACD, the mixture is transferred to a sterile 50 ml conical plastic centrifuge tube and it is centrifuged at 200 *g* for 15 minutes. The upper two-thirds of the platelet rich plasma (PRP) is transferred using disposable plastic pipettes into a 15 ml conical plastic tube and centrifuged at 2000 *g* for 10 minutes, following which the platelet poor plasma (PPP) is carefully removed and saved. The cell button is then resuspended in 4 ml of the ACD solution, recentrifuged and the supernatant removed. The platelets are then resuspended in the 4 ml of ACD solution containing the ^{111}In-oxine complex and this is then incubated at room temperature for 20 minutes. Following incubation. the suspension is centrifuged at 2000 *g* for 15 minutes and the supernatant removed. The platelets are then resuspended in 4 ml of the PPP which was previously saved and incubated for a further 7 minutes. The suspension is centrifuged again at 2000 *g* for 15 minutes, the supernatant removed, and the cells resuspended in a further 5 ml of PPP, after which they are ready for injection.

Indium labelling of transferrin

Both [113m]In and ^{111}In are normally obtained as indium chloride and the quantity with the required activity (about 5 MBq for a plasma volume determination with [113m]In or about 75 MBq for bone marrow imaging with ^{111}In) is incubated with 10 ml of the patient's plasma at 37°C for 30 minutes after which it is ready for injection. Some workers have preferred to use purified solutions of transferrin instead of the patient's plasma and a solution of 100 µg/ml in 5 per cent sodium citrate has been found suitable. Unfortunately, not all the indium is firmly bound to the transferrin by incubation, and it is necessary to remove the loosely-bound indium from the labelled plasma before injection. This can be achieved by passing the labelled plasma through a sterilized ion-exchange column (Chelex 100 from Biorad is suitable; Keeling, 1971). Most workers have not done this, and this possibly explains the relatively poor correlation between [113m]In-labelled transferrin and ^{125}I-labelled HSA in plasma volume measurements reported by Wootton (1976).

Iron labelling of transferrin

The radionuclides of iron are normally obtained as ferric citrate made isotonic with sodium chloride but some are supplied as ferric chloride in HCl solution. In the latter case it is necessary to neutralize and buffer the solution before use by the addition of a suitable quantity of sodium citrate. For ferrokinetic studies the unsaturated iron binding capacity of the patient's plasma should be determined before labelling is commenced, and, if this is less than 0.5 mg/ℓ, normal donor plasma should be used instead. For non-imaging studies about 200 kBq of radio-iron should be added to 10 ml of plasma and incubated under sterile conditions for 30 minutes at 37°C, while for imaging studies with [52]Fe about 7.5 MBq should be used. It has been suggested (Cavill, 1971) that any non-transferrin-bound iron should be removed before injection by passing the labelled plasma through a sterile anion exchange resin column (eg Amberlite IRS 400 Cl⁻ from Koch-Light Limited). The majority of workers, however, do not consider this necessary for the simple ferrokinetic tests although it is certainly required for the more sophisticated analyses of the clearance curve (*see* Chapter 5).

References

CAVILL, I. (1971) The preparation of^{59}Fe labelled transferrin for ferrokinetic studies. *Journal of Clinical Pathology*, **24**, 472–474

DACIE, J. V. and LEWIS, S. M. (1975). *Practical Haematology*, 5th Edn. pp. 471–472. Edinburgh and London: Churchill Livingstone

FERRANT, A., LEWIS, S. M., and SZUR, L. (1974). The elution of ^{99m}Tc from red cells and its effect on red cell volume measurement. *Journal of Clinical Pathology*, **27**, 983–985

HEATON, W. A., DAVIS, H. H., WELCH, M. J., CARLA, J. M., JOIST, J. H., SHERMAN, L. A., and SIEGEL, B. A. (1979) Indium-111: A new radionuclide label for studying human platelet kinetics. *British Journal of Haematology*, **42**, 613–622

ICSH (1971) Recommended methods for radioisotope red cell survival studies. *British Journal of Haematology*, **21**, 241–250

ICSH (1977) Recommended methods for radioisotope platelet survival studies. *Blood*, **50**, 1137–1144

ICSH (1980) Recommended method for radioisotope red cell survival studies. *British Journal of Haematology*, **45**, 659–666

ICSH (1980) Recommended methods for measurement of red cell and plasma volume. *Journal of Nuclear Medicine*, **21**, 793–800

JONES, J. and MOLLISON, P. L. (1978) A simple and efficient method of labelling red cells with ^{99m}Tc for determination of red cell volume. *British Journal of Haematology*, **38**, 141–148

KEELING, D. H. (1971). *A study of 113m Indium labelled pharmaceuticals*. pp. 32–33. MSc Thesis: University of London

WOOTTON, R. (1976) The limitations of ^{113m}In for plasma volume measurement. *British Journal of Radiology*, **49**, 427–429

A derivation of Dornhorst's equation for cell survival

It is assumed that cells are destroyed by two mechanisms; first, a random destruction mechanism which has the same probability of destroying cells regardless of their age and secondly, a mechanism due to senescence which destroys cells only when they reach a certain age. Either or both mechanisms may be present in any population of cells and it is assumed that a random population of cells are labelled and thus the labelled cells are representative of cells of all ages in the circulation.

Calculation of the number of labelled cells surviving in the circulation at a time (t) after injection

The rate of destruction of cells due to the random destruction mechanism alone will be exponential so that we can write:

$$\mathrm{d}N/\mathrm{d}t \propto \mathrm{e}^{-kt}$$

where N is the number of labelled cells and k is the rate of random destruction.

The rate of destruction due to the senescence mechanism alone will be constant so that for this mechanism we can write:

$$\mathrm{d}N/\mathrm{d}t \propto C$$

where C is a constant.

Combining these two functions gives:

$$\mathrm{d}N/\mathrm{d}t \propto C \cdot \mathrm{e}^{-kt}$$

and integrating this expression we obtain:

$$N(t) = \frac{-C}{k}\mathrm{e}^{-kt} + B$$

Where $N(t)$ is the number surviving at time t and B is a constant. This can be rewritten as in equation (1)

$$N(t) = A e^{-kt} + B \tag{1}$$

Where A is a constant and is equal to $-C/k$. There are two boundary conditions that can be applied; first, if T is the age at which death occurs due to senescence then $N(T) = 0$, and secondly, when $t = 0$, $N(t)$ is equal to the number of labelled cells injected (N_0). It follows from the first condition that $A e^{-kT} + B = 0$ and from the second condition that $N_0 = A + B$. From these two equations expressions for A and B can be obtained in terms of k, T and N_0 and substituted in equation (1) which is then given by the form in equation (2).

$$N(t) = N_0 (e^{-kt} - e^{-kT})/(1 - e^{-kT}) \tag{2}$$

This is the general equation derived by Dornhorst.

For the special case when there is no death due to senescence we can make T equal to an infinite time, then e^{-kT} becomes equal to zero and $N(t)$ becomes equal to $N_0 e^{-kt}$, so that the survival curve is purely exponential.

For the special case when there is no random destruction, as k approaches zero so e^{-kt} approaches $1 - kt$ and then $N(t)$ becomes equal to $N_0 (1 - t/T)$ so that the survival curve is linear.

Calculation of the mean cell life

The MCL is given by the reciprocal of the initial rate of destruction as shown in equation (3).

$$\text{MCL} = - \left[\frac{d}{dt} \left(\frac{N(t)}{N_0} \right)_{t=0} \right]^{-1} \tag{3}$$

For random destruction alone $N(t)/N_0 = e^{-kt}$ so that the MCL is equal to $1/k$.

For death due to senescence alone $N(t)/N_0 = 1 - t/T$ so that the MCL is equal to T.

For death due to both mechanisms:

$$\frac{N(t)}{N_0} = \frac{(e^{-kt} - e^{-kT})}{(1 - e^{-kT})}$$

and the differential of this at $t = 0$ is equal to $-k/(1 - e^{-kT})$ so that the MCL is equal to $(1 - e^{-kT})/k$.

Calibration of imaging equipment

In order to obtain the calibration factor that relates the number of detected γ-rays making up an image of an organ to the quantity of the radionuclide in the organ, a 'phantom' containing a known amount of the radionuclide is imaged in a water bath under various conditions. The procedure used for calibrating rectilinear scanners and hybrid imaging devices is described first and this is followed by a simple discussion of the procedure for calibrating the other imaging devices.

For imaging the phantom the following procedure is used:

1. The quantity of the radionuclide to be used (A) is measured by counting it in a standard volume in a syringe fixed in a jig mid-way between the two detectors of the imaging equipment.
2. The contents of the syringe are then transferred into a suitable phantom (eg a 250 ml polythene bottle) and this is positioned in the centre of a tank filled with water to a depth of 250 mm.
3. The phantom in the water tank is imaged and the scanning speed (S) is noted.

The calibration factor (K) is then given by the expression $K = A/(NS)$ where N is the number of detected gamma rays making up the image of the phantom. This factor is only accurate for a patient thickness of 250 mm and further images are therefore recorded with different depths of water so that corrections may be made for the absorption and scatter of γ-rays in patients of different thickness. Measurements may also be made at the same time to check the independence of the detector sensitivity to the position of the phantom in the tank by imaging it at different positions in the water bath.

For imaging the organ in the patient the following procedure is used:

1. The syringe containing the radionuclide to be injected is filled to the standard volume and counted in the jig in the same way

as for the phantom scan. It is also recounted after injection so that the exact quantity injected can be determined.

2. An image of the organ is then obtained and the scanning speed noted. It is important to ensure that the area imaged covers a larger area than just the organ of interest, so that a background region may be used in order to correct for any of the radionuclide still circulating in the blood or taken up in the surrounding tissue. With short-lived radionuclides it is also necessary to note the time interval between the measurement of the syringe before injection and the recording of the image, so that a correction can be made for the decay of the radionuclide.

3. After imaging, but before the patient is moved, the profile of the patient should be determined at the position of the organ being imaged.

The calculation of the quantity of the radionuclide in the organ may be done either manually or by computer. Most imaging equipment is now connected to some form of computer and the algorithm required closely follows the manual calculation. Since this is simple to follow it is described below.

An area is delineated around the image of the organ that should be slightly larger than the apparent size of the organ. Close to this 'region of interest' an additional area is marked out in order to obtain a background region—ideally the body profile over the background region should be similar to that over the organ. The region of interest over the organ is then subdivided into areas of approximately equal body thickness, and the number of γ-rays forming the image in each of these sub-regions and the background region are determined. The number of 'background' counts expected in each sub-region is determined from the relative areas of each sub-region and the background region. This background is then subtracted from the number of detected γ-rays in each sub-region and the resulting number is multiplied by the calibration factor corresponding to the thickness of the patient at this position. The corrected number of γ-rays in each of the sub-regions is then summed and the percentage of the quantity of the radionuclide injected in the organ is calculated using the expression, $P = 100\ NS/A$, where P is the percentage uptake in the organ, N is the total corrected number of γ-rays forming the image of the organ, S is the speed at which the organ was scanned and A is the quantity of the radionuclide injected. The calculation of N, the corrected number of γ-rays, for a simple image recorded on a traditional rectilinear scanner is shown in *Figure A3.1* and *Table A3.1*.

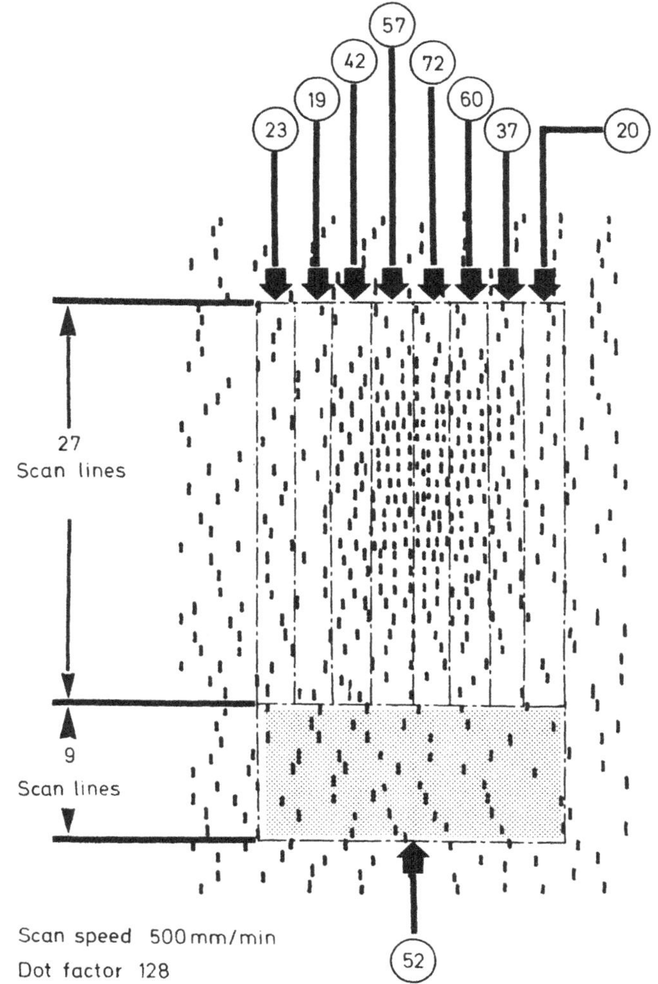

Figure A3.1 The division of a rectilinear scan of an organ into regions of interest and the summation of the detected counts in each sub-region

If a gamma camera is used to record the images it is necessary to obtain calibration factors for both the depth of the organ in the body and also for the separation of the body surface from the gamma camera face. This again is most simply achieved by imaging a phantom in a water bath, the water level and phantom

Table A3.1 Calculation of the total corrected number of γ-rays forming the image shown in Figure A3.1

Background counts/strip $= (52/8) \times (27/9) = 19.5$

Dots/strip	Background subtracted	Thickness (mm)	Calibration factor	'Corrected' dots/strip
23	3.5	206	0.82	2.9
19	—	208	0.82	—
42	22.5	207	0.82	18.5
57	37.5	204	0.81	30.4
72	52.5	199	0.78	41.0
60	40.5	190	0.74	30.0
37	17.5	178	0.69	12.1
20	0.5	164	0.64	0.3

Total number of 'corrected' dots in image $= 135.2$
Dot factor $= 128$
Total corrected number of γ-rays forming the image $= 135.2 \times 128 = 17\,306$

position then being altered until all combinations have been covered. When imaging the patient it is necessary to take orthogonal views of the organ of interest (ie two images at 90° to each other) so that the organ's depth beneath the skin and its distance from the collimator face of the detector can be determined. In addition it may be necessary to correct the image for non-uniform sensitivity across the field of view of the gamma camera. Once this has been done the calculation of the organ uptake is performed as described above except that the imaging time is substituted for the scanning speed. Because of the way gamma camera images are normally displayed, it is essential to have the gamma camera interfaced to a computer to record the imaging information.

With tomographic scanning equipment, provided that the algorithm used for the reconstruction of the cross-sectional images is accurate, each cross-sectional image of the organ should accurately show the distribution of the radionuclide and no correction for the depth of the organ is necessary. The problem of determining the total amount of the radionuclide in the organ then becomes one of determining the effective thickness of each cross-sectional image since this will govern how many different cross-sectional images need to be summed in order to obtain the total number of γ-rays forming the total organ image. This is best achieved by, as before, imaging phantoms in a water bath—the number of cross-sectional images that need to be summed is equal to the total length of the organ divided by the effective thickness of each slice. The summed images should be equally spaced along the length of the organ. The percentage uptake is then simply obtained by comparing the number of γ-rays forming the image with the imaging time and the quantity injected in the same way as for the other imaging devices.

Radioassay techniques

The techniques of radioassay have had a very large impact on the measurement of low concentrations of biochemically active substances in the body over the 20 years since the methods were first described. Although most of the initial work and major advances were made in the field of endocrinology their application has now spread to many diverse fields and most haematology laboratories are now involved with at least some radioassay work. The range of substances of haematological interest which can be measured by these methods is, however, very large and because of this only the underlying principles of the technique are discussed here and not the details of individual assays.

The methods used can be broadly divided into two types (Ekins, 1979), the so-called 'limited reagent' or 'saturation analysis' methods and the 'excess reagent' methods. Of these, the first group currently comprise most of the widely-used radioassays but the second group has certain theoretical advantages and is gradually becoming more popular.

The basic principles of both types of method are the same and depend upon interaction between the test substance and a reagent following which the quantity of the test substance present is inferred from an observation of the extent of the reaction. In the first group the reagent used is present in a limited and carefully defined quantity which is less than the amount required for reaction with all the test substance. Thus, by looking at the ratio of the amount of test substance which has undergone reaction to the amount which has not undergone reaction it is possible to obtain a measure of the quantity originally present. In the second group the reagent is present in greater quantities than is required for complete reaction with all the test substance and in this case it is the ratio of the two fractions of the reagent (the reacted and the unreacted) after the reaction has taken place which provides the measure of the quantity of the test substance present. The two methods are summarized in equations (1) and (2) below.

Limited reagent method

$$\begin{matrix} \text{Test substance} \\ + \\ \text{limited reagent} \end{matrix} \quad \rightarrow \quad \begin{matrix} \text{test substance complex} \\ + \\ \text{residual test substance} \end{matrix} \qquad (1)$$

Excess reagent method

$$\begin{matrix} \text{Test substance} \\ + \\ \text{excess reagent} \end{matrix} \quad \rightarrow \quad \begin{matrix} \text{'converted' reagent} \\ + \\ \text{residual reagent} \end{matrix} \qquad (2)$$

The basic principle of the limited reagent method is, perhaps, most simply illustrated with the aid of a simple analogy. Following Ekins (1974), if we have a jug of water and a glass and the jug holds an unknown volume of water which is greater than the capacity of the glass, then, when the jug is emptied into the glass some, of the water will overflow out of the glass. The amount that overflows and the ratio of the amount that overflows to the amount in the glass depend on the volume originally in the jug. Similarly, in the excess reagent method the capacity of the glass is made greater than the unknown volume in the jug, and the ratio of the volume in the glass to the unused volume of the glass or to the total volume of the glass is measured.

In practice, identical quantities of the reagent are added to the test samples and to standard samples containing known amounts of the test substance. Then, by comparing the ratios of the two fractions present after reaction, it is possible to determine the quantity of the substance in the unknown samples.

In order to utilize the technique, means must be found for separating the two fractions and measuring the distribution of the substance between them. In limited reagent methods the latter is most commonly achieved by adding a small identical amount of radionuclide-labelled test substance to each sample before adding the reagent and assuming that this is partitioned in an identical way to the unlabelled test substance in the sample and, in the excess reagent methods, the reagent is similarly labelled with a radionuclide tracer. A measurement of the quantity of radioactivity in each fraction of each sample will then be directly proportional to the quantity of the test substance or reagent in each fraction. The reagent must be as specific as possible for the test substance because if it also binds to other substances, it becomes necessary to purify the original sample. The separation of the two fractions is commonly achieved by the use of protein-coated charcoal to absorb the unreacted material so that the final separation of the two fractions is simply achieved by the physical separation, by centrifugation, of the charcoal from the rest of the sample. Other methods are also used as appropriate to the chemical form of the materials to be separated.

Normally the distribution of the radionuclide is expressed in one of the following two ways: the ratio of the radioactivity in the two fractions, either bound/unbound or unbound/bound; or as the percentage of the total radioactivity in the sample which is either bound or unbound. The results from the standard samples are then plotted as a function of the amount or concentration of the unlabelled substance in them in order to obtain a standard curve. The quantity of the substance in the unknown samples is then obtained by reading off the corresponding values from the standard curve.

The shape of the standard curve varies with the assay being done and also with the exact way in which it is done. For example, in the assay of vitamin B_{12} in serum it is often found that an approximately linear relationship is obtained if the data is plotted as a log–logit as shown in *Figure A4.1* (logit x = log (x/1–x)), while in the assay of folate a simple linear plot is usually found to be best (*see Figure A4.2*).

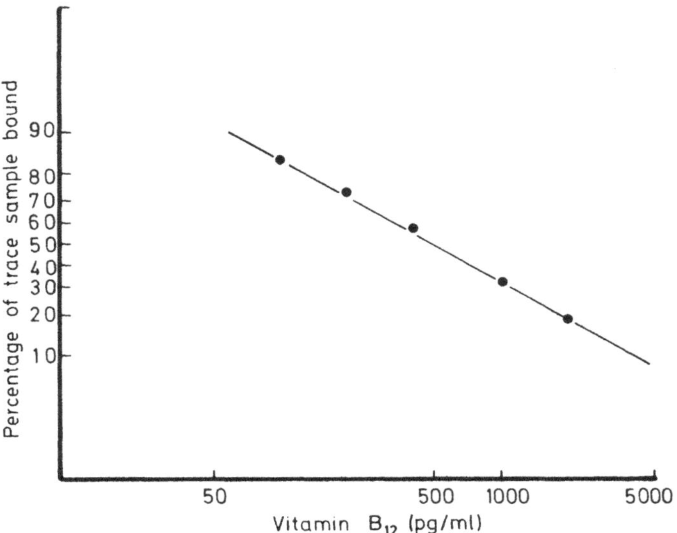

Figure A4.1 A typical standard curve obtained for an assay of vitamin B_{12} (log–logit plot)

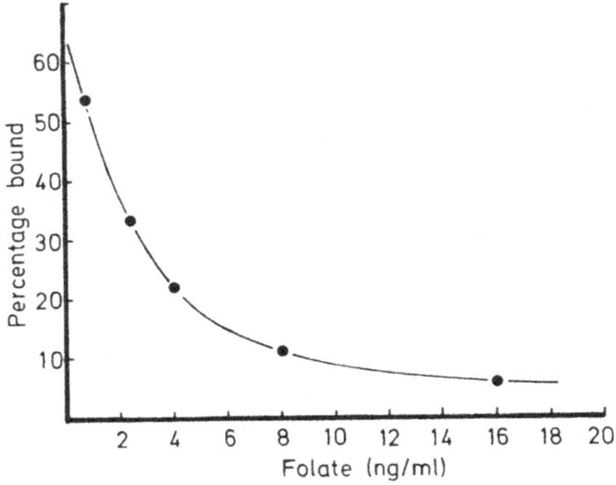

Figure A4.2 A typical standard curve obtained for an assay of folic acid (linear plot)

There are a great number of different assays of haematological interest available commercially in 'kit' form. Some of these, such as those used for vitamin B_{12}, folate, ferritin and iron binding capacity, replace old and well-established microbiological assays and have the advantage of being much simpler to use. But others, such as that for β-thromboglobulin, which provides a measure of platelet function (Ludlam *et al.*, 1975), and that for fibrinogen degradation products, which provides a measure of intravascular coagulation and fibrinolysis (Gordon *et al.*, 1975), allow completely new measurements to be made. It is, however, usually necessary for each laboratory to establish its own range of normal values before using an assay for clinical purposes and care must be taken to ensure that the assay is reproducible both within itself and from assay to assay. Finally, it is often worthwhile to refer to the original papers that describe a particular assay in order to see what problems were encountered with it during development—this will often provide insight into any difficulties that may be experienced. These references are usually quoted in the manufacturers' literature.

References

EKINS, R. P. (1974) Radioimmunoassay and saturation analysis: basic principles and theory. *British Medical Bulletin*, **30**, 3–11

EKINS, R. P. (1979) Radioassay methods. In *Proceedings of the Second International Symposium on Radiopharmaceuticals.* p. 219–239. New York: Society of Nuclear Medicine

GORDON, Y. B., MARTIN, M. J., LANDON, J. and CHARD, T. (1975). The development of radioimmunoassays for fibrinogen degradation products: fragments D and E. *British Journal of Haematology,* **29,** 109–119

LUDLAM, C. A., MOORE, S., BOLTON, A. E., PEPPER, D. S. and CASH, J. D. (1975) The release of a human platelet specific protein measured by radio immunoassay. *Thrombosis Research,* **6,** 543–548

Index

Made in the USA
Monee, IL
07 July 2026

56550261R00085